Robert F Thuma

A New Necessary Science. A study for Teaching or

Self-Improvement

The Grace of Man

Robert F Thuma

A New Necessary Science. A study for Teaching or Self-Improvement
The Grace of Man

ISBN/EAN: 9783337163761

Printed in Europe, USA, Canada, Australia, Japan

Cover: Foto ©berggeist007 / pixelio.de

More available books at **www.hansebooks.com**

A New Necessary Science.
A Study for Teaching or
Self Improvement.

The Grace of Man.

—By—
ROBERT F. THUMA.

ILLUSTRATED.

Pittsburgh, Pa., U. S. A.
The Myers & Shinkle Company, Printers
1897.

Copyrighted 1897,
By ROBERT F. THUMA.

Dedicated

To the Pupils who in the past have zealously studied the precepts of this work and by their progress in Physical and Spiritual Development have inspired me with the thought of putting in print the Ethics of the " Grace of Man."

THE highest desire any reasonable man can cherish, or the highest prerogative he may possibly claim is to become perfect. Desire results from attraction, attraction results from the separation of two substances, analogous in their essences and properties. This was well known to the Ancients, although claimed by modern science; modern science indeed, which deals in nothing more, as a rule, than observations, and the classification thereof, of extempore phenomena, the causes of which it knows naught of; whereas, I claim that science in its purest and highest view, is the self knowledge of the fundamental laws of Nature, and is in consequence a spiritual science, based upon the knowledge of our inner selves.

Man is composed essentially of seven principles each dependent in a way one upon the other; clearly demonstrating the alleged modern axiom, "action and reaction are equal." We have but three principles of man to deal with directly in the following treatise; but to the student, who desires a full comprehension, let him investigate his other principles, and learn the significance of the six pointed star.

INDEX.

	PAGE.
Frontispiece, "Ashes of Roses."	
Instructions,	ix
Analysis of Subjects,	xii
Synopsis—The Thuma Study of Grace,	1

Gravity, ... 7
   Seven Exercises for the Control of Centre of Gravity—Why, ... 11
   Derivations—Exercises for the Control of Centre of Gravity, ... 13
   Seven Laws for the Control of Gravity 16

Corrective Exercises of the Physical Body,
                      Series I.—(From Base to Waist Line,) 17
   Seven Corrective Exercises for the Physical Body, ... 21
   Derivations—Corrective Exercises for the Physical Body, ... 23
   Seven Laws for the Physical Corrections of the Body, (Series I).. 26

Corrective Exercises of the Physical Body,
                      Series II.—(Waist to Neck Inclusive), 29
   Seven Corrective Exercises for the Physical Body, ... 33
   Derivations—Corrective Exercises for the Physical Body, ... 35
   Seven Laws for the Physical Corrections of the Body, (Series II), 38

INDEX.—Continued.

PAGE.

Corrective Exercises of the Physical Body,
        Series III.—(Head and Arms)........ 41
  Seven Corrective Exercises for the Physical Body,............. 45
  Derivations—Corrective Exercises for the Physical Body,....... 47
  Seven Laws for the Physical Corrections of the Body, (Series III), 50

—◦—

The Anatomy of the Elements of Man—The Trinity,...... 53
  The Soul, The Physical Body and Grace,..................... 55
  The Location of the Elements in the Physical Body,........... 57
  Sub-division—the Head,..................................... 58
  Sub-division—the Trunk, ... 59
  Sub-division—the Extremities,............................... 60
  The Main Forces,... 61

—◦—

The Study of Curved Lines,........................... 63
  Derivations,.. 65
  The Alphabet,.. 66
  The Seven Arm Positions,................................... 67
  Circles—Mental,.. 68
  Seven Circle Exercises,..................................... 70
  Wave Lines—Emotional,..................................... 74
  Seven Wave Line Exercises,................................. 76
  Spiral Movements—Power,................................... 82
  Seven Spiral Exercises,..................................... 86
  Seven Laws for the Curved Lines,........................... 93

INDEX.—Continued.

	PAGE
Flexibility	95
Seven Flexing Exercises,	98
Seven Relaxation Exercises,	100
Seven Falling Exercises,	102

Attitudes 105
   Straight Attitudes, ... 108
   Arch Attitudes, ... 110
   Bendings, ... 112
   Straight (Knee Bend) Attitudes, ... 114
   Arch (Knee Bend) Attitudes, ... 116
   Kneeling Attitudes, ... 118
   Reclining Attitudes, ... 118
   Lying Attitudes, ... 118
   Poise, ... 120

Force 121
   Reserve Power, ... 124
   Exercises for the Cultivation of Reserve Power, ... 125
   Spiritual Force, ... 126
   Aphorisms to Cultivate Spiritual Force, ... 127
   Physical Power, ... 128
   Seven Walking Exercises, ... 129
   Will Power, ... 130
   Aphorisms to Cultivate Will Power, ... 131
   Personal Magnetism 132
   Exercises for the Cultivation of Personal Magnetism, ... 133
   Hypnotism 134

INDEX.—Continued.

                                                                                     PAGE.

Poise, Opposition and Harmony..................... 135
    Exemplification,... 137
    Harmony,... 138
    Laws of Opposition and Harmony, Showing the Harmony of the
        Whole,.. 139
    Analysis of Attitudes, Showing Harmony and Opposition,....... 146

The Mind... 151
    Axioms for Developing the Mind,........................... 153
    Inspiration,.. 154
    Ideality,... 155
    The Study of the Soul,..................................... 157
    To Cultivate the Soul,..................................... 160
    The Soul Responds,.. 161
    Duality,... 163

Studies in Grace.. 165
    The Blending Pictures—Descriptive,........................ 166
    The Visions, or the Play of the Emotions—Descriptive,........ 167
    A "Solo"—Taken from Exercises in this Book—Descriptive,.... 168
    The Blending Pictures—Illustrated Explanation, ... 170
    The Visions, or the Play of Emotions—Illustrated Explanation,. 176
    A "Solo"—Illustrated Explanation,......................... 180
    Music—Viola Gavotte,..................................... 182

    Appendix,... 184

INSTRUCTIONS.

INSTRUCTIONS.

The primary object to be borne in mind is that the philosophy of this science is

> THE EXPRESSION OF THE SOUL,
> THROUGH THE MIND,
> BY MOTION OF THE BODY.

First, the body is perfected with physical corrective exercises for the body.

Second, the mind is placed in touch with movements of the body, by expressing mental intentions and feelings with movements of the body.

Third, these mental expressions with body movements, are intensified, until the soul expression (or personal magnetism) is developed and the soul is reached.

KEYNOTE.

The keynote upon which this work is founded is the
> TRINITY,
> THREE IN ONE.

Each part of the Trinity having a sub-division of
> SEVEN PARTS—(no more, no less),

Each separate part of the seven having
> THREE HARMONIES.

SUBJECTS.

In the Table of Analysis of Subjects, which follows, will be found Seven Trinities of Subjects. Each subject being complete in itself, but is a part of the other two.

GRAVITY.

Gravity is the foundation of Poise and Grace and is one of the first studies in this work to be mastered. It will bear an analysis of three divisions or seven derivations and contains twenty-one exercises.

PHYSICAL BODY.

For the physical perfection the body is divided into three distinct series of seven parts each, (no more, no less), and these are given three distinct exercises for the perfection of each one of these parts, namely:

>SERIES   I.—From Base to Waist Line.
>SERIES II.—Waist to Neck, inclusive.
>SERIES III.—Head and Arms.

Each series containing seven parts consisting of twenty-one exercises.

CURVED LINES OR GRACE OF MOVEMENTS.

The study of Grace or Curved Lines is the result or reflection of perfect physical corrections.

FORCE.

The development of the forces within us, starts with the first exercise, and will be found applicable or adjuncts of all the studies. The finish of the Grace of Man is the perfection of the Physical body, and the development of the Forces within us. The end being Personal Magnetism.

DUALITY.

The study of Duality is carried on throughout this work, although in esoteric form. In the duality of matter the force or element that connects it, creates the Trinity.

COURSE OF STUDY.

The subjects given in the analysis (next page,) are placed in this work as they should be taken up. Each study is complete under separate headings, starting with easy construction for No. 1, and increasing to construction No. 7, none of which are difficult to execute. The student or teacher should read the book first and then master each part complete.

APPENDIX.

The Appendix given on last page of this book, shows how to impart this study to the pupil, or class, and may be followed either by the teacher or student. That is, taking the simple constructions of No. 1, of all the subjects, and adding until the book is mastered. The number of lessons necessary should be left to the teacher.

Analysis of Subjects.

The Trinity of Man
is the Soul, the Mind, and the Body.

I. Gravity.   Flexibility.   Force.
   (Mental.)   (Emotional.)   (Power.)

II. The Physical Corrections of the Body.
   Series I.   Series II.   Series III.
   (Physical.)   (Emotional.)   (Mental.)
   Seven parts to each Series, showing Perfection, Relation, Harmony.

III. The Anatomy of the Elements of Man.
   The Trinity.
   The Mental.   The Emotional.   The Physical.
   Showing the Trinity in every part of the Physical body.

IV. Attitudes.   Opposition.   Harmony.
   (Mental.)   (Emotional.)   (Strength.)

V. Curved Lines.   Wave Lines.   Spiral Lines.
   (Mental.)   (Emotional.)   (Strength.)

VI. The Mind.   Inspiration.   Ideality.

VII. The Soul.   Personal Magnetism.   Duality.

SYNOPSIS.

THE THUMA STUDY OF GRACE.

BY ROBERT F. THUMA.

"THUMA ZITHRA" is the translation of this motto. *Thuma* meaning soul, and *Zithra*, grace, or the "grace of soul," and it is the seal of the writer. To be graceful of body, the mind must be cultivated by thought and study that pertains to our subject, and through that unseen instrument, the mind, we can reach the soul. Therefore to be truly graceful, we must have grace of soul and leave the movements of the body to be but the reflection of our innermost being.

Did you ever see a graceful person who had not corresponding traits in mind and soul? On this premise I shall attempt to show why it lies within the power of every one to be as graceful as he could desire.

When I speak of grace, I mean the glorifying kind, the beautiful that elevates and purifies, and there is no grander, stronger gift than that to be desired.

My idea is that, "The Grace of God" means the synthesis of all power, beauty, goodness and charity. The very essence of all that relates to the higher world, is embodied in these words, "The Grace of God." On this basis I have founded my study, and the grace of man will bear an analysis which is readily comprehended by us mortals.

A potent law of life is, that "habit is second nature." (That much abused word "habit.") No other word, will exactly express the idea. That is, if you form a bad habit or a good one, it becomes a part of your very existence. Speaking of mortals, what are some of the unwritten elements that make some men and women tower above all others in that unseen strength which we feel in their presence, but can not explain? Why are there great statesmen, actors, preachers, and men who carry us by the mere force of their personality?

They possess the Grace of Man in some form or other. Can these forces be cultivated? Yes. If we are patterned after our Creator, then what is in the spiritual is also in the flesh; you may enlarge a muscle, you may cultivate the mind, you can give the soul within more freedom. To commence this study we must start with the body, and cultivate every muscle in the body by correcting the imperfections. You will find in this fine instrument, the body, an adjustment and harmony that cannot be duplicated. Take up the study of the beautiful, it feeds the eyes,—the windows of the soul. The next is the study of grace of movement of every part of the body, this includes the harmony of one muscle with the other.

If the student has practiced corrective exercises, he will find his movements a matter of grace and ease. As yet we have only prepared the body, in training as it were, for a more complete end of this necessary study, to be a fully developed person and to have a complete existence.

After going through enough practice to satisfy the body, put the Mind, that storehouse of intelligence through which passes the link which connects the soul with the body, in action with these movements of grace of body. Think of beautiful subjects, and use heavenly music, for is not music "the language of angels;" conceive thoughts of this nature and express them through the movements of the body. To illustrate, we mention a few creations of the author: "The Visions of One on High," "Angels at Play," "In a Garden Fair," "Death—A Play of the Emotions," "Castles in the Air," "The Artist's Dream," all of which relates to the subject in hand.

A practice of this kind will develop an unseen force—Personal Magnetism,—which you can feel and which will intensify with nursing and proper environment. Personal Magnetism is created by intensifying the thought of the mind until the soul expression is reached. We can then give our soul freedom to glorify, to live and prosper. Can you not see how a man's soul is narrowed? To study and practice the beautiful in movements with the body trained, fostering the mental forces on the same plane, will surely bring out the soul expression, that will carry all before it.

Having reached this point you will then have no trouble to reverse the order, that is to say, start at the soul, transmit through the mind to the body or parts thereof. Have your soul start the action and express it by the personal magnetism or soul expression in physical form of action of the body, all of which are subservient to that instrument, the mind.

Now this theory must be thoroughly understood. It has been a life study of the writer, and is entirely practicable. These teachings are not only for the reader, but there is also a regular course of practice to accomplish these results.

We have in times past learned "Harmony of Poise" through the Delsartean Theory, and by many students it has been gathered that you can trace a system of infinite harmony in all things, man, animals, nature, and in the elements, and as we go upward the spiral cycle enlarges to that magnitude that is far beyond the mind of mortal. It is written and stands to-day, God the Father, Son and Holy Ghost, the Trinity, the Three in One. What is the Grace of God? Love, Wisdom, Power. In man the corresponding elements are the emotional, the mental and the physical. Man is composed of these elements. The perfect being is one that is endowed with an amount of these elements, harmoniously balanced.

The perverted man may be all mental and physical and no emotion, or all mental and emotional and lack physical, and there are as many different men and women as there are possible combinations and tints, as it were, with these different elements. This theory will apply to all colors, to all arts, to all things; suffice it to say that to a student of this lore, the elements have their location in the different parts of the body, and in each part of the body. This is true of the plant, the earth and every atom of life.

The first work of this new study then is the correction of imperfections, such as lack of muscular activity of the limbs, the correct carriage of the head on the shoulders, the position of the feet, the straightening of the back, the discontinuance of one muscle doing the work of another. For instance, the neck muscles under the ears strengthening the carriage of the head instead of making the vertebra muscle in the back do all the work. The head is the seat of our mentality; as we carry it, so will the entire body try to adjust itself to a harmonious carriage.

These are only hints of the study, which complete, are for every part of the body—the toes, heel, knee, thigh, waist, chest, shoulders, elbow, wrist, hand, fingers, back, spinal column, neck, head, etc. Three months study of an hour or two each week and a fair knowledge of the work can be gained, to say nothing of the benefits derived.

The study of the beautiful, the curved lines, the wave lines, the spiral movements, follow in regular order to complete the work. Among the other studies embraced herein are Reserve Power, Personal Magnetism, Inspiration, Gravity, Flexibility and Strength, Poise and all subjects that relate to THIS STUDY OF GRACE,
which is

THE EXPRESSION OF THE SOUL,
THROUGH THE MIND,
BY MOTION OF THE BODY.

GRAVITY.

GRAVITY.

The Control of the Centre of Gravity of Body.

All bodies, the earth, a ship, or a pencil have a centre of gravity. To be graceful the student should be able to balance the body, have the muscles under control, that the body may assume any graceful attitude easily balanced. The first object to acquire to be graceful, is to be able to balance the body. I do not think a Delsartean or any one could be graceful without this part of our study.

The method of finding the centre of gravity of any object is worth knowing. In a solid ball, perfectly round, the centre of gravity would naturally be exactly in the centre. Of a long shaped article of even weight it would be in the middle, or of an unevenly shaped or weighted body it must be nearer one end. Thus a ball rolling down a hill would eventually put the gravity in the centre, and for instance, the earth not being evenly weighted, the gravity is not exactly in the centre of the mass. To ascertain the centre of gravity of an unevenly weighted object, *i. e.* a stone, suspend the same from any point and pass a plumb line from point of suspension to base (perpendicular,) then re-hang the stone from any other point and repeat line; at the intersection of these two lines would be the centre of gravity, or the point of its balance. Now by this explanation we can understand what is meant by gravity.

Gravity of body consists of reserve power of the mind, buoyancy or natural adjustment of the body, and equilibrium, in any attitude or position.

GRAVITY.

Consists of 1. Reserve Power—(Mental)—no waste of vital force.

2. Buoyancy —(Emotional)—natural adjustment of muscles.

3. Equilibrium —(Physical)—strength of poise.

DERIVATIONS.

1. Control of centre of gravity.

2. Buoyancy.

3. Gravity of diagonal lines.

4. Perpendicular uniformity.

5. Propelling power.

6. Beyond the base line.

7. Mental forces.

I.
PENDULUM.

Step to right on right foot and swing left in front of right, now repeat, step to left on left foot and swing right in front of left. Hands on hips. Continue to alternate, like a pendulum. Count 1, 2, 3, to each swing. (waltz time.) Do 16 times.

II.
PIROUETTE.

Step to right with right foot, cross left toe over right, turning completely around while raised on toes. Finish in a closed position. Repeat to left. Hands on hips. Count 1, 2, 3, 4, 5, 6, (waltz time,) to each turn. Do continuous, 16 times.

III.
MERCURY BALANCE.

Step to right and assume this attitude, balance thus down and up on one foot. 8 times. (waltz time, 8 bars.) Repeat to left. Do 4 times.

IV.
THE ADVANCE.

Step forward on right and balance on that foot by drawing the left up to right in the back (raised.) Count 1, 2, 3, now repeat by stepping back on left and balance on that foot by drawing the right up to left in front, (raised.) Count 1, 2, 3. (Waltz time.) Hands on hips. Do this continuous, 16 times.

V.
BUOYANCY.

Slide right foot out in front and then quickly draw it back, at same time slide left foot out and draw it back. Hands on hips. Do this continuous. Count 1 & 2 & (march time.) Do 16 times.

VI.
PIROUETTE AND BENDING.

Pirouette (II.) to left and then face to right and slide left foot back and bend to left. (per cut.) Slowly recover and repeat to left. Hands on hips. Continue to alternate. Count 1, 2, 3, 4, 5, 6, to each Pirouette and 1, 2, 3, 4, 5, 6, to each Bending. (Waltz time.) Do 8 times.

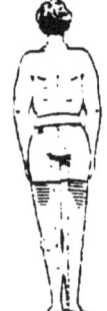

VII.
THE HEAD.

See law of gravity for line from neck to heel. Practice foregoing exercises observing the gravity positions of the head. The head taking the rhythm.

SEVEN EXERCISES FOR CONTROL OF CENTRE OF GRAVITY—WHY.

	TITLES.	GRAVITY	THE MIND.	HARMONY
I.	PENDULUM.	Perpendicular Uniformity.	Rythm.	Regularity of Motion.
II.	PIROUETTE.	Control of Centre.	Reserve Power.	Balance of Forces. No vital waste, unnecessary motion or force lost.
III.	MERCURY BALANCE.	Diagonal Lines.	Mental Control of Gravity leaving the will decide incline.	Head and Acting Foot.
IV.	ADVANCE.	Propelling Power.	Rythm.	Regularity of Motion.
V.	BUOYANCY.	Grace of Muscles.	Induce Elasticity.	Natural Adjustment.
VI.	PIROUETTE BENDINGS.	Beyond the Base Line.	Reserve Power.	Suppleness Curve.
VII.	HEAD.	Mental Forces.	Control of Force.	Intent.

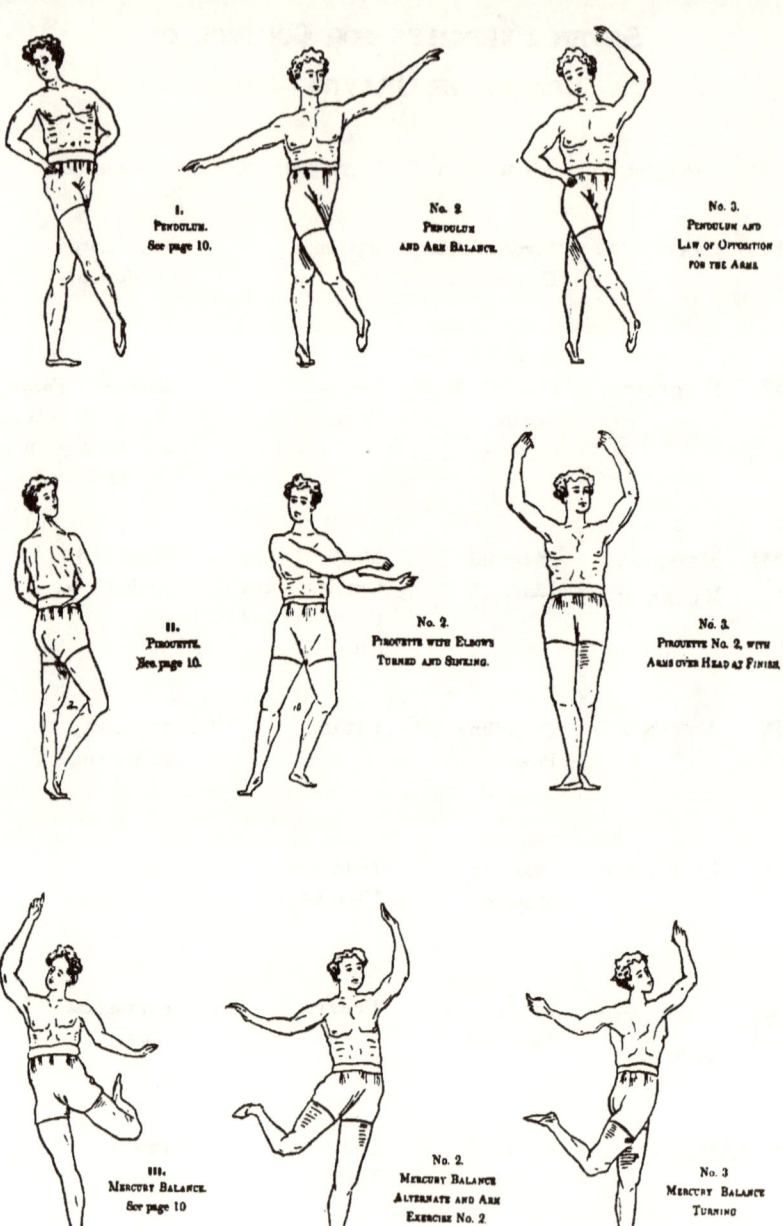

EXERCISES FOR CONTROL OF CENTRE OF GRAVITY.

DERIVATIONS.

I.

1. PENDULUM.—See explanation page 10.
2. PENDULUM AND ARM BALANCE.—Same as No. 1, with this addition—extend arms and sway the body with the swing of the foot. Do 16 times. Waltz time.
3. PENDULUM WITH LAW OF OPPOSITION FOR THE ARMS.—Same as No. 1, with this addition—Raise right arm over head, when you swing left foot in front, and left arm over head when you swing right foot. Continue to alternate 16 times. Waltz time.

II.

1. PIROUETTE.—See explanation page 10.
2. PIROUETTE, WITH ELBOW TURNED AND SINKING.—Same as No. 1, with this addition—when turning to right crook right elbow and extend left arm. This should be done while the right foot is being extended. Dissolve this arm position as you are turning and crook left elbow and extend right arm. The hands are now in position, turn left. The sinking (or bending and raising) of the knee is done as you assume the closed position at the finish. Count 1, 2, 3, for Pirouette, and 1, 2, 3 for sinking and rising. Do 16 times. Waltz time.
3. PIROUETTE No. 2, WITH ARMS OVER HEAD AT FINISH.—Do No. 2 movement complete, at turning crook elbow, at sinking hands on hips, and at finish of feet to closed position, place hands over head, as in cut. Do alternate 16 times. Waltz time.

III.

1. MERCURY BALANCE.—See explanation page 10.
2. MERCURY BALANCE ALTERNATE, AND ARM EXERCISE No. 2.—Step to right, assume Mercury Balance attitude, count 1, 2, 3, sinking down and up in this position, count 1, 2, 3. Repeat to left, count 1, 2, 3–1, 2, 3. To place arms in position (see attitude)—use Arm Exercise No. 2 as you change from right to left. Continue to alternate. Do 16 times. Waltz time.
   NOTE.—For arm movement, see study of curved lines.
3. MERCURY BALANCE TURNING.—Assume Mercury Balance to right and pivot (turning) on right foot ¼ turn to each count of 1, 2, 3. Repeat to left. Do twice to right and twice to left. Waltz time.

ADDITIONAL.—Note for 2. Do Pirouette No. 2, followed by Mercury Balance No. 2, to right. Repeat complete to left, is a very pretty combination.

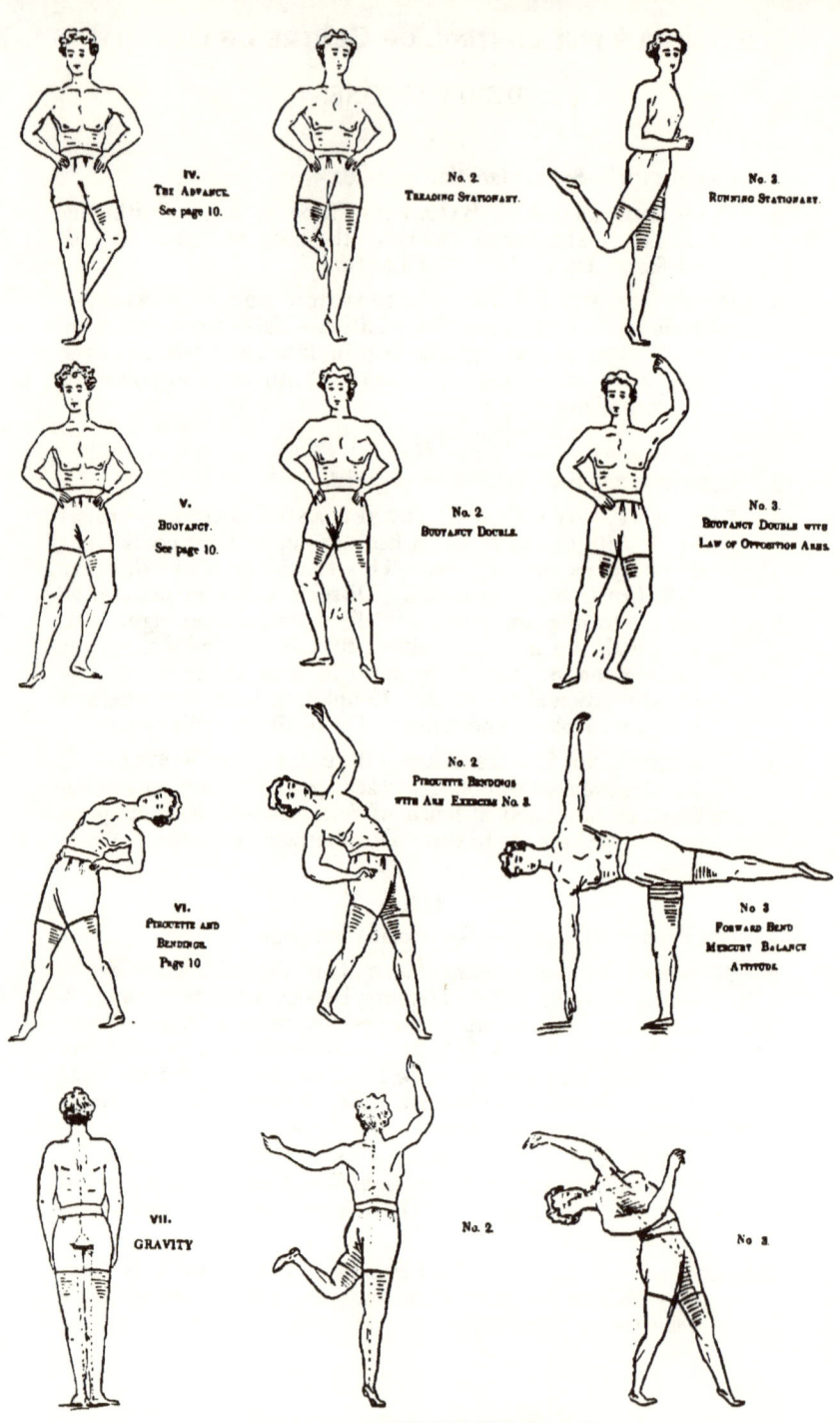

LAWS OF GRAVITY —See page 16

IV.

1. THE ADVANCE.—See explanation page 10.
2. TREADING STATIONARY.—Mark time with the feet, increasing the elevation of the foot (the toe of one foot reaching the knee of the other.) March time, 32 bars.
3. RUNNING STATIONARY.—Hands closed, mouth closed. At each tread swing the foot behind (well up) as if running but remaining on the same spot. This requires a strong buoyancy or balancing of the body at each tread. Continue until body is adjusted to a natural swing. March time, 32 bars.

V.

1. BUOYANCY.—See explanation page 10.
2. BUOYANCY DOUBLE.—Double the original movement on each foot. That is, slide right foot out in front, then quickly draw it back, repeat with right—at same time slide left foot out and draw it quickly back, repeat with left foot. When doubling the movement hop on standing foot. Do alternate 16 times. March time.
3. BUOYANCY DOUBLE, WITH LAW OF OPPOSITION FOR ARMS.—Do Buoyancy No. 2 and when the right foot is acting raise the left arm over head, and when left foot is acting, raise right arm over head. Do alternate 16 times. March time.

VI.

1. PIROUETTE AND BENDINGS.—See explanation page 10.
2. PIROUETTE BENDINGS WITH ARM MOVEMENT NO. 3.—Same as No. 1, with this addition—crook the elbow during the Pirouette and add Arm Movement No. 3 during the bend, as in cut. 6 counts for Pirouette and 6 counts for Bendings. Do alternate 8 times. Waltz time.
3. FORWARD BEND. MERCURY BALANCE ATTITUDE.—Assume Mercury Balance attitude on right foot. Bend slowly forward and have high (right) hand touch floor. Recover and repeat with left foot and left hand. Count 1, 2, 3, 4, 5, 6, 7, 8, for each bend. Do 8 times. Gavotte music.

VII.

HEAD.—The Head indicates and finishes the Harmony and Gravity of any position. (See Laws of Harmony.)

As you carry your Head so will you carry your whole body.—(See Laws of Head.)

The Head is the heaviest separate instrument of the body. It is the God Head of our body. It expresses something in any attitude. (See Laws of Head.)

Practice the laws of Gravity of Head with all the foregoing exercises.

LAWS FOR CONTROL OF GRAVITY.

1. A line drawn from centre of back of neck down and diverging at base of spine to the centre of each heel would illustrate a person in a perfect state of gravity while standing.

2. When on one foot, line from neck to heel of standing foot, as in Mercury Balance and Pirouette, shows regulation of gravity.

3. In Bending, a line drawn from centre of the base directly up is the seat of the control of gravity, which requires an opposition position : from this point at any given place, may be another subjective control, which again requires an opposition at that point.*

4. Law of Turning or Pirouette is to step (intent) and turn in that direction.

5. To illustrate the law of the control of gravity. As the frame work of muscles are to the braces in a bridge or butterfly wing—they are or become rigid when assuming a poise and the change of attitude of the body necessitates the constant change of gravity, and therefore becomes mechanical.

6. The Head is always in opposition to the highest hand, when hand is above the line of vision.—(See Laws of Opposition.)

7. There must be an opposition of weight from centre of gravity and from subjective centres and this is changed as often as the attitude changes, which constitutes the centre of gravity or equilibrium.

*NOTE.—See the laws of Opposition and Cut of No. 3 Gravity, page 14.

Corrective Exercises of the Physical Body.

Series I.

From Base to Waist Line.

Physical Corrections of the Body.

(From Base to Waist Line—Series I.)

1. The Foot.

The base on which the body rests is the feet and ankles and should have early training in this study ; then we will be sure of a good foundation. The Toes, Heels and Ankles, should have considerable attention. So closely allied are the different exercises that they should be practised in combination to retain their relationship. Here I desire to mention that of all the different parts of the body that will be studied, the Head being the seat of the intellect or from whence the intelligence to direct and the intent of action is formulated, is the most important.

2. The Ankles.

The Ankles should be of the same use to the feet that the wrists are to the hand by flexing the ankle, the muscles are stretched, after which in combination with other corrective exercises (the Lock, Toe Position, etc.) they will assist control of Foot and force the Foot to grow or turn out. The harmony of Knee Joint and Hip must be applied as the latter is the seat from whence the Feet or Limbs should turn out.

3. The Heel.

The close relation of the Heel to Hip muscle is here mentioned, as the Hip is the source and the Heel is the end. From the Heel are continued other muscles which control the foot.

4. The Knee.

The Knee bears the same relation to the body as the spring to the bed of a carriage. As we progress the jar of the entire body is passed off at the bending of the Knees. Thus, to jump and keep the Knees rigid, will jar the different muscles, even to the hair on your head, but to bend the knee as you alight will relieve all jolts that the body would otherwise experience.

5. THE IMPORTANCE OF THE HIP.

The Hip is the physical element of the leg, (see divisions of the anatomy of elements of man) and gives the power to the lower extremities. As the foot is the mental, the knee is the emotional, thus in order to turn out the toes (or foot) the toes should turn out—giving the intent, then the knee-joint, and then Hip (or Power) should be applied and will naturally train the foot to turn out. A practice of these exercises will easily cure so called pigeon-toed people.

6. THE THIGH.

The Thigh should be strong and reliant and is of great advantage when so, in all attitudes where the legs are braced, and crossed in bending.

7. THE LEG.

The Leg is the machinery of our locomotion and that denotes force. The best exercises which are given are those that relate to the individual control of each limb. Under the subject of walking, given elsewhere more information may be had.

I.
TOE POSITION EXERCISE.

Extend right foot in front, toe touching floor. Tap 4 times. Repeat with left foot. Hands on hips. Leg always stretched. See laws for toe. Do alternate 8 times. Schottische time.

II.
ANKLE TWIST.

Rise on right toe as in cut, now step down on, right foot, and quickly raise on left toe. Repeat with left toe. Hands on hips. Twist heel well out front each time. Count 1 & 2 & for right and left foot. Do continuous 32 times. March time.

III.
HEEL TWIST.

Step on right heel, (see cut) count 1. quickly change weight to ball of left foot 2, then to ball of right foot 3. Repeat with left heel. Hands on hips. Twist the entire foot when on heel from in to out. Do alternate 16 times. Polka time.

IV.
KNEE SPRING.

Spring and kneel as per cut, (front view to right, count 2 bars of Polka music. Repeat to left. Hands on hips. Use elasticity of knee in the spring to relieve the jar in alighting, and bend backward when kneeling. Do alternate 8 times. Polka time.

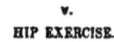

V.
HIP EXERCISE.

Face to right, hop on right foot and raise left, count 1, hop again on right foot, at same time twist left out and face left, count 2, (as in cut.) Now hop on left and repeat with twisting right out, count 3, 4. Hands on hips. Do alternate 16 times. Schottische time.

VI.
THIGH BRACE.

Cross right foot over left (weight on both feet,) now turn the body completely round, (on ball of foot,) but do not release your foot-hold as originally crossed, (per cut.) Now return body to place. Hands on hips or extended. Do 8 times. Repeat by crossing left foot over right. Do 8 times. Count 1, 2, 3, 4, 5, 6, to each turn. Waltz time.

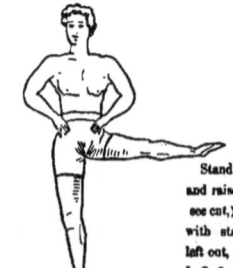

VII.
LIMB EXTENSION.

Stand on left foot and right toe. Extend and raise right foot straight out at side, see cut,) repeat slowly 8 times. Repeat with standing on right foot and extend left out, 8 times. Hands on hips. Count 1, 2, 3, 4, 5, 6, 7, 8, for each extension and finish. Gavotte music.

Seven Corrective Exercises of the Physical Body.

(From Base to Waist Line.—Series I.)

	TITLES.	CORRECT.	RELATION.	HARMONY.
I.	Toe Position.	To point toe.	Stretching front muscles of leg from waist to toe.	Toe to knee to hip.
II.	Ankle Twist.	Turning out of feet.	Pliability of, and use of base.	Heel and toes. Control of feet.
III.	Heel Twist.	Inclination of heel to standing foot—Turning out of foot.	Stretching back muscles of leg. Outward turn of knee, thigh and hip.	Hip to knee joint to heel.
IV.	Knee Spring.	Springiness, Activity of the lower limbs.	Bouyancy, Activity of hip muscles and groin.	Shoulders to carry the head properly.
V.	Hip Exercise.	Turning out of extremities.	Groins, Power of the leg.	Transmitting of force to these seven parts.
VI.	Thigh Brace.	Support. Strength.	Brace in Poise, Gravity.	Firmness of Carriage in Poise.
VII.	Limb Extension.	Individual control of lower limbs.	Control of Gravity in Poise.	Mental determination. Directed control of the seven parts of this series.

1.
TOE POSITION EXERCISE.
See page 20.

No. 2.
TOE POSITION
AND ARM EXTENSION.

No. 3.
TOE POSITION WITH
LAW OF OPPOSITION
OF THE ARMS.

II.
ANKLE TWIST.
See page 20.

No. 2.
ANKLE TWIST
ADVANCE AND RETIRE.

No. 3.
ANKLE TWIST
WITH "LOCK" OF HANDS,
ADVANCE AND RETIRE.

III.
HEEL TWIST.
See page 20.

No. 2.
PLAIN
TOE AND HEEL.

No. 3.
TOE, HEEL, TOE AND
PAUSE.

CORRECTIVE EXERCISES FOR THE PHYSICAL BODY.

(From Base to Waist Line.—Series I.)

DERIVATIONS.

I.

1. TOE POSITION EXERCISE.—See explanation page 20.
2. TOE POSITION AND ARM EXTENSION.—Same as No. 1, with this addition—with arms extended, when executing the four taps with each foot. Do 16 times. Schottische time.
3. TOE POSITION WITH LAW OF OPPOSITION OF THE ARMS.—Same as No. 1, with this addition—raise left arm over head, when executing the 4 taps with right foot, and right arm over head when doing the same with left foot. Continue to alternate 16 times. Schottische time.

II.

1. ANKLE TWIST.—See explanation page 20.
2. ANKLE TWIST, ADVANCE AND RETIRE.—Same as No. 1, with this addition—move forward (advance) executing this Ankle Twist 16 times, then move backward (retire) executing the Ankle Twist 16 times. Hands on hips. Repeat 4 times. March time.
3. ANKLE TWIST WITH "LOCK."—Lock the arms or hands up the back (per cut,) assume an erect carriage, with arms thus locked and executing Ankle Twist, advance 16 times, and retire 16 times. Repeat 4 times. March time.

III.

1. HEEL TWIST.—See explanation page 20.
2. PLAIN TOE AND HEEL.—Hop on right foot and touch left toe on floor, (near right heel,) count 1. Hop again on right foot and touch left heel on floor (near right toe,) count 2. Hop on right foot and touch left toe on floor (near right heel,) count 3. Hop on right foot and touch left heel on floor (near right toe,) count 4. Now change to left from last position, hop on left and touch right toe to left heel, etc., count 1, 2, 3, 4. No stop for change of right to left. Do alternate 8 times. Schottische time.
3. TOE, HEEL, TOE AND PAUSE.—Similar to Plain Toe and Heel, only omit movement of count 4, and pause instead, at movement of count 3. Thus hop on right foot and touch left toe on floor, (near right heel,) count 1. Hop again on right foot and touch left heel on floor, (near right toe,) count 2. Hop again on right foot and touch left toe on floor, (near right heel,) count 3. Pause in this position, count 4. Now change to left, count 1, 2, 3, (pause 4.) No stop for change of right to left. Do alternate 16 times. Schottische time.

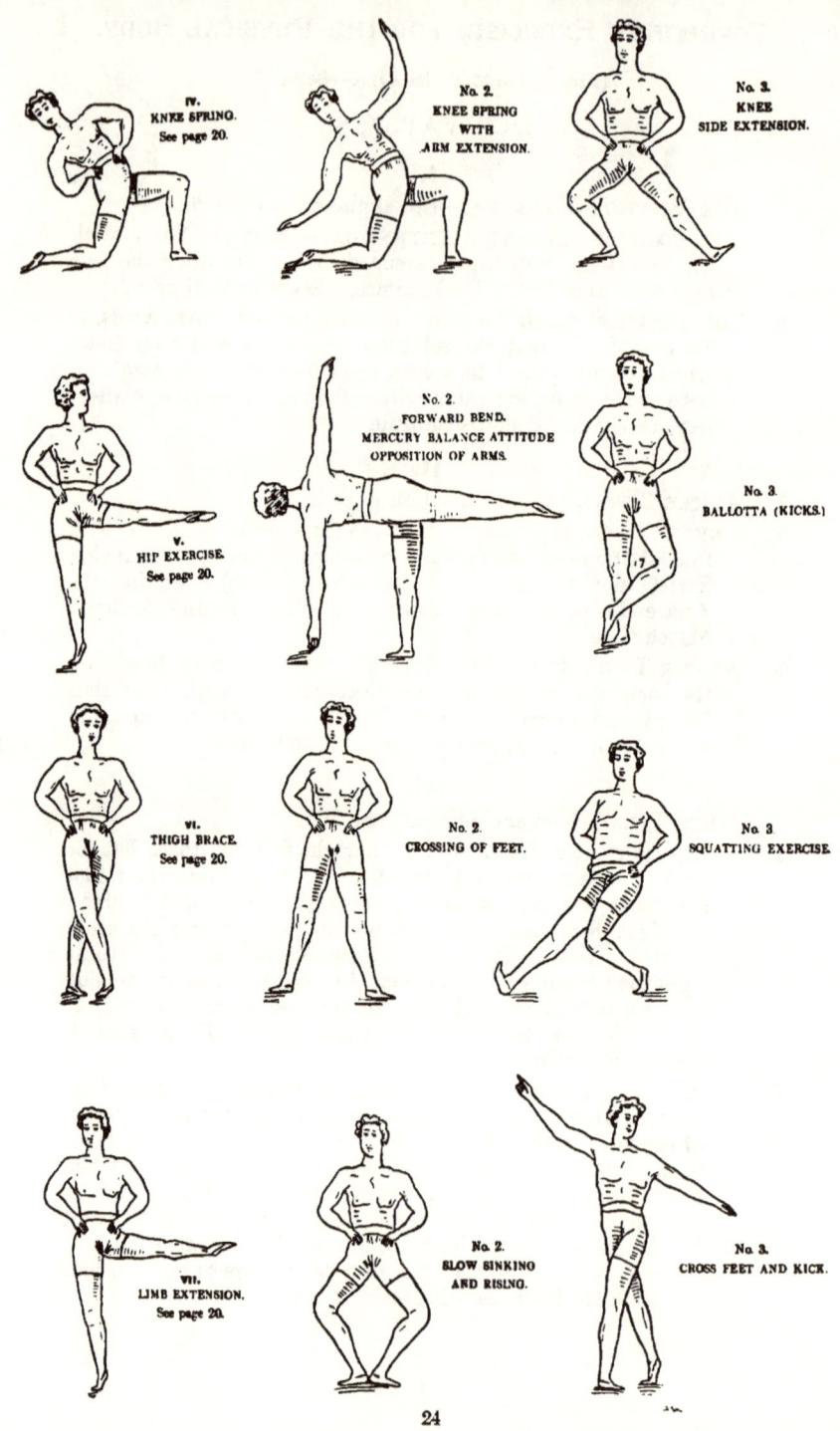

IV.
1. KNEE SPRING.—See explanation page 20.
2. KNEE SPRING WITH ARM EXTENSION.—Same as No. 1, with this addition—extend arms as in cut, bend back when kneeling down, and pass arms through Arm Exercise No. 1, as you change from right to left. Do alternate 8 times. Polka time.

NOTE.—For arm movements, see study of curved lines.

3. KNEE SIDE EXTENSION.—Spring and assume attitude (per cut,) right side extension; spring and assume left side extension; execute from right to left continuous. Count 1, 2, for each extension. Do 16 times. Schóttische time.

V.
1. HIP EXERCISE.—See explanation page 20.
2. FORWARD BEND MERCURY BALANCE ATTITUDE AND OPPOSITION OF ARMS.—Assume Mercury Balance attitude, (cut No. III. page 10,) and bend slowly forward, touching floor with fingers. Use opposition of arms thus, standing on right foot touch floor with left hand; now change attitude to left foot and touch floor with right hand, count 1, 2, 3, 4, 5, 6, 7, 8, for each bend and recover. Do 4 times. Gavotte music.
3. BALLOTTA (KICKS.)—Ballotta means to kick out. Hop on right (same time cross left toe over right foot,) count 1; hop on right foot (same time kick left foot out at side,) count 2; now change to left. From last position, hop on left foot (same time cross right toe over left foot, per cut,) count 3; hop on left foot (at same time kick right foot out at side,) count 4. No stop for change of right to left. Hands on hips. Do alternate 16 times. Schottische time.

VI.
1. THIGH BRACE.—See explanation page 20.
2. CROSSING OF FEET.—Hop off both feet and assume a crossed position of the feet, (count 1 &); hop off both feet and assume position (per cut,) (count 2 &). Now cross and re-cross feet, before, behind and before, rapidly, count 1, 2, 3. Total 2 bars of music. Hands on hips. Do 8 times. Polka time.
3. SQUATTING EXERCISE.—Assume squatting or sitting position. In this position extend right foot, count 1, 2, 3, (per cut,) draw it back and extend left foot, count 1, 2, 3. Hands on hips. Do alternate 8 times. Polka time.

VII.
1. LIMB EXTENSION.—See explanation page 20.
2. SLOWLY SINKING AND RISING.—Slowly sink down (per cut,) then assume an upright position until up on toes. Hands on hips. Count 1, 2, 3, 4, 5, 6, 7, 8, to each sinking and rising. Do 16 times. Gavotte time.
3. CROSS FEET AND KICK.—Cross right foot over left (put weight on right,) count 1; hop on right, and kick left foot out at side, count 2. Now as left is in air, from this position cross left over right foot, (put weight on left,) count 3; hop on left and kick right out at side, count 4. Hands on hips or extended. Do continuous 16 times. Schottische time.

SEVEN LAWS FOR THE PHYSICAL CORRECTIONS OF THE BODY.

(From Base to Waist Line—Series I.)

THE FOOT.—(THE TOE.)

The toe must always be pointed downward and outward with the heel drawn up. In walking, the additional harmony of ankle must be used. Thus in advancing, the toe strikes the floor first, and in retiring the toe leaves the floor last. The toe is the mental element of the foot.

THE ANKLE.

The pliability of the ankle must be applied to the walking movement and to all motions of the foot individually, to give it the expression intended.

THE HEEL.

In any open position, in any attitude or pose the heel should incline toward the standing foot. The toe pointed in opposition out and downward—this law will apply to the foot in any position. The heel is the mental element of the hip.

THE KNEE.

The knee is the seat of locomotion or movement of the body; viz: "sinking and rising" and no progressive physical movement of any description is not embraced therein. In descending the knee should always bend and return to an upright position with the knee cap turned out. Buoyancy is generated at the knee and upon it greatly depends the natural adjustment of the muscles in the "bouyancy of the body." Strength and perfection of the springing activity is developed here.

The Hip.

The Hip should act as the physical power of the lower extremities, and that power should be transmitted to the knee, thence to the foot and toes ; viz: in turning the foot outward in walking, and in all movements of force of the body this is demonstrated.

The Thigh.

The Thigh or lever from hips to knees, should be the support or brace (between) as it were, to support the Gravity in Poise, Crossed Position, and Bendings. To avoid shakiness of Poise this muscle must be·well controled.

The Leg.

The Leg in repose should be perfectly straight, the Limb Extension and other leg exercises given here are combination studies of Change of Gravity and Reserve Power, and serve to control muscles of the limbs as an enity. They are for the individual control of each limb and these exercises should be followed with this intention.

CORRECTIVE EXERCISES OF THE PHYSICAL BODY.

Series II.

WAIST TO NECK INCLUSIVE.

Physical Corrections of the Body.

(Waist to Neck, inclusive.—Series II.)

1. The Waist.

The Waist is the supposed centre of control of gravity, and the muscles leading from it are the largest and strongest at this point. They are used to constitute the opposition balance at the seat of gravity. They are as the first spreading branches from the parent stem of a tree, from which the smaller branches shoot forth.

2. The Trunk.

The Trunk is the store-house and machine shop of some of the finest subjective machinery. Here we find the stomach, to which is fed the fuel to generate and run this machinery. The Heart is the engine of life, the Kidneys the refinery and the Lungs the store-house, etc. This machinery should be controlled by the mind. The Soul being the inner atmosphere pervading all as the air is "life to exist," so is the soul to every part of the body. Without air vegetation ceases, and when the soul leaves, the body ceases in a short time to even be a body.

3. The Spinal Column.

The Spinal Column is the main stay of the trunk, and its flexibility is only to the extent of its natural inclination, and varies with different pupils. It should never bend beyond its natural scope. After the natural bend of the Spinal Column, the neck muscles and head at one end, and the groins and knee muscles at the other, should be brought into use to complete the arching.

4. The Abdomen and Stomach.

These parts, in regard to grace, are subjective to the waist line, and are known as the physical elements of the trunk, (see anatomy of the elements of man), and should never protrude outward or inward to an extreme degree. To rest these two parts on one hip is a common habit of standing and should be immediately corrected.

5. The Chest.

Beauty of carriage of the body greatly depends upon the position of the chest. It contains the lungs, which are the mental element (see the anatomy of the elements of man) of the trunk, and should therefore (mental) be given prominence by observing the law of opposition of chest and chin, (see opposition laws). The chest can only hold this point of prominence when there is a harmonized action of the shoulder blades.

Breathing. The Groins.

Proper breathing is important to life, and the best known method, taken from the Italian school, is given here only as a corrective exercise of the functions of the body. That is, inhale to store the air below the stomach until the diaphram is extended, and exhale without the heaving of the chest, using the vocal channel as a stationary canal. In a case of shortness of breath, stretching of groins in harmony with the diaphragm should be used, as they relate to the chest in breathing.

6. The Back.

The back to be perfect should be straight. To do this the chest should have its proper position, the shoulder blades should not protrude, and the head should be "aplomb" on the neck or spinal column. To reach this state of perfection, use Elbow Twist exercise, which will flex the shoulder blades, and use such muscle exercises that relate to flexibility of the shoulder blades and a proper carriage of the head. When these muscles become pliable, use the "Lock" exercise, locking the muscles of the back in a smooth position until habit retains them so.

7. The Neck.

The neck muscles require considerable attention for the proper carriage of the head. The head is the seat of intelligence, and as we carry our head, in any slight shade or change of poise so will we carry the body. We should never loose sight of this fact. The cords or muscles under the ears and in front, should be strengthened to hold back the head, to strengthen the vertebra at the back of the neck. In cases where one is hollow chested, the head hanging forward, and has developed a long neck, this may be changed by assigning the front muscles their duty, and giving the head its proper adjustment.

I.
WAIST SPANISH CIRCLE.

Kneel down and assume this attitude. Circle arms around to right (fingers nearly on floor,) carry them over head (bending back,) lower them to left side and circle to starting attitude, making one continuous motion. Circle to right 16 bars and to the left 16 bars. 2 or 4 bars to each circle. Waltz time.

II.
TRUNK TWIST.

Stand firm. Hands on hips. Twist right side of trunk to front sharp, count 1 &. Now twist left side to front, count 1 &. Then twist trunk quickly 3 times from right, left, right side, count 1, 2, 3. Repeat starting left. Do 4 times. Polka time, (slow.)

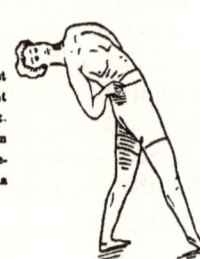

III.
SPINAL COLUMN NATURAL

Do Thigh Brace Turn and Bend, thus: Thigh Brace Turn is step to side on left foot, cross right well over left, (weight on both feet,) remain in this position, now turn body (on ball of foot) to left until face front. Bend back to left as in cut. Slide left foot and recover, repeat same Thigh Brace Turn, to right and bend back right. Count 2 bars to turn, and 2 bars to bend and recover. Do alternate 8 times. Waltz time.

IV.
ABDOMEN AND STOMACH CIRCLE.

Assume this attitude, facing diagonal front, (right or left.) Feet in stationary position, execute a circle (by bending knees) with abdomen and stomach thus; sink down, move same forward, rise to starting attitude. Do with right foot extended 8 times, and left foot extended 8 times. 2 or 4 bars for each circle. Waltz time.

V.
CHEST AND SHOULDER BLADES.

For Chest Expansion and pliability of Shoulder Blades. Clap hands in front (per cut,) then behind. Walking movement (marking time stationary,) one step to each bar of music. Clap hands with each step on first beat in bar. Do 32 times. Waltz or March time.

VI.
THE "LOCK" FOR BACK.

Lock hands in back thus, palms together, assume upright position. If carriage of chest head and back are not perfect, pull lock up higher. In this locked position do Ankle Twist, (Series I.) thus rise on right toe, bend well out front, step down on right foot and immediately raise left toe. Do continuous alternate 32 times. March time. Observe perfect carriage of body.

VII.
NECK EXERCISE.

Point right foot at side. Hands on hips. Rise on toes (spring motion) and by stretching neck muscles, and look over your back, so you can see heel of left foot, (as per cut.) Draw the right elbow and back in. Recover and repeat to left side. Count 2 bars polka to each side, when proficient 1 bar. Do 16 times. Polka time.

Seven Corrective Exercises of the Physical Body.

(From Waist to Neck Inclusive. Series II.)

	TITLES.	CORRECT.	RELATION.	HARMONY.
I.	Waist Spanish Circle.	Control of Centre. Freedom of Extremities.	Seat of the strong muscles of physical power.	Strength of muscles in relation to opposition and Control of Gravity.
II.	Trunk Twist.	Ease of Carriage of Body.	Groin muscles. Gravity of bending beyond base line.	Reserve power subjective control of movement from the mind.
III.	Spinal Column Natural.	Cultivate its Natural Bend.	Control of the trunk and erect carriage.	Natural expansion of this instrument and all muscles of the back.
IV.	Abdomen and Stomach Circle.	Its complete subjective control in all motions of the body.	Seat of our physical life or animal nature.	Source of supply of vital power to the physical elements, thence to spiritual.
V.	The Chest and Shoulder Blades.	Enlarge the lungs and correct carriage of trunk.	As the brain to the head, so are the lungs to the trunk.	Ease of erect carriage. Chin, chest and shoulder blades.
VI.	The Lock for Back.	To straighten the back.	Shoulders, chest and head.	Shoulder blades, spinal column. Chest expansion and proper carriage of body.
VII.	The Neck Exercise.	Proper carriage of head by strengthening neck muscles.	Muscles of head to carriage and motion of head.	As a swivel or ball-bearing joint for head.

1.
WAIST
SPANISH CIRCLE.
See page 32.

No. 2.
SPANISH CIRCLE
STANDING.

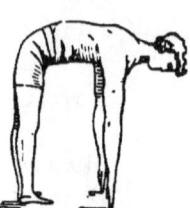

No. 3.
FORWARD BEND
TRACE ¼ CIRCLE
WITH FINGERS ON FLOOR.

II.
TRUNK TWIST.
See page 32.

No. 2.
THE PA de BASQUE.

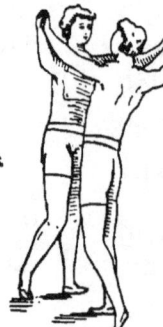

No. 3.
THE SIDE BEND
IN PAIRS.

III.
SPINAL COLUMN
NATURAL.
See page 32.

No. 2.
SPINAL NATURAL
WITH ARM
EXERCISE No. 3.

No. 3.
DRAW BACK
BENDS.

CORRECTIVE EXERCISES FOR THE PHYSICAL BODY.

(From Waist to Neck inclusive.—Series II.)

DERIVATIONS.

I.

1. WAIST SPANISH CIRCLE.—See explanation page 32.
2. SPANISH CIRCLE STANDING.—Stand firm with right foot extended and do same waist circle with arms and waist as No. 1. Do 16 times to right. Extend left foot and do 16 times to left. Waltz time.
3. FORWARD BEND, TRACE ½ CIRCLE WITH FINGERS ON FLOOR.—Start with hands over head, now bend forward (do not bend knee) and touch the floor with fingers of both hands, trace ½ circle to right in this position and around to left, assume upright position. Repeat 4 times. Gavotte music, 32 bars.

NOTE FOR No. 1.—Additional exercise is to kneel and swing hands forward, and then bending backward, swing hands over head.

II.

1. TRUNK TWIST.—See explanation page 32.
2. THE PA DE BASQUE.—Step to right with right foot, cross left in front and balance, count 1 for right foot, 2 for left foot, 3 for balance, (at balance release left foot.) Repeat to left. Twist right side of trunk well around when starting with right foot, and twist left side of trunk to left when starting with left foot. Hands on hips. Do alternate 16 times. Mazurka time.
3. THE SIDE BEND IN PAIRS.—Assume attitude as per cut in pairs. Sway from right to left. Balance on right foot when you sway to right, and balance on left foot when you sway to left. Bend trunk well to each side. Count 2 bars waltz time to each bend. Do 16 times.

III.

1. SPINAL COLUMN NATURAL.—See explanation page 32.
2. SPINAL NATURAL WITH ARM EXERCISE No. 3.—Same as No. 1, with this addition, cross elbow when executing Pirouette, and Arm Exercise No. 3 for the bend, as per cut. Count 1, 2, 3, 4, 5, 6, to each Pirouette, and the same for bending. Do alternate 8 times. Waltz time.

NOTE.—For Arm Exercise No. 3, see Study of Curved Lines.

3. DRAW BACK BENDS.—Step right foot behind, count 1 and assume side view, (Neck Exercise) as per cut, count 2. Stretch neck over so that you can see right heel, right arm over head. Hold this attitude, count 3, 4. Now place left foot behind, left arm over head, and repeat in opposition, count 1, 2, and hold 3, 4. Do 8 times. Schottische time.

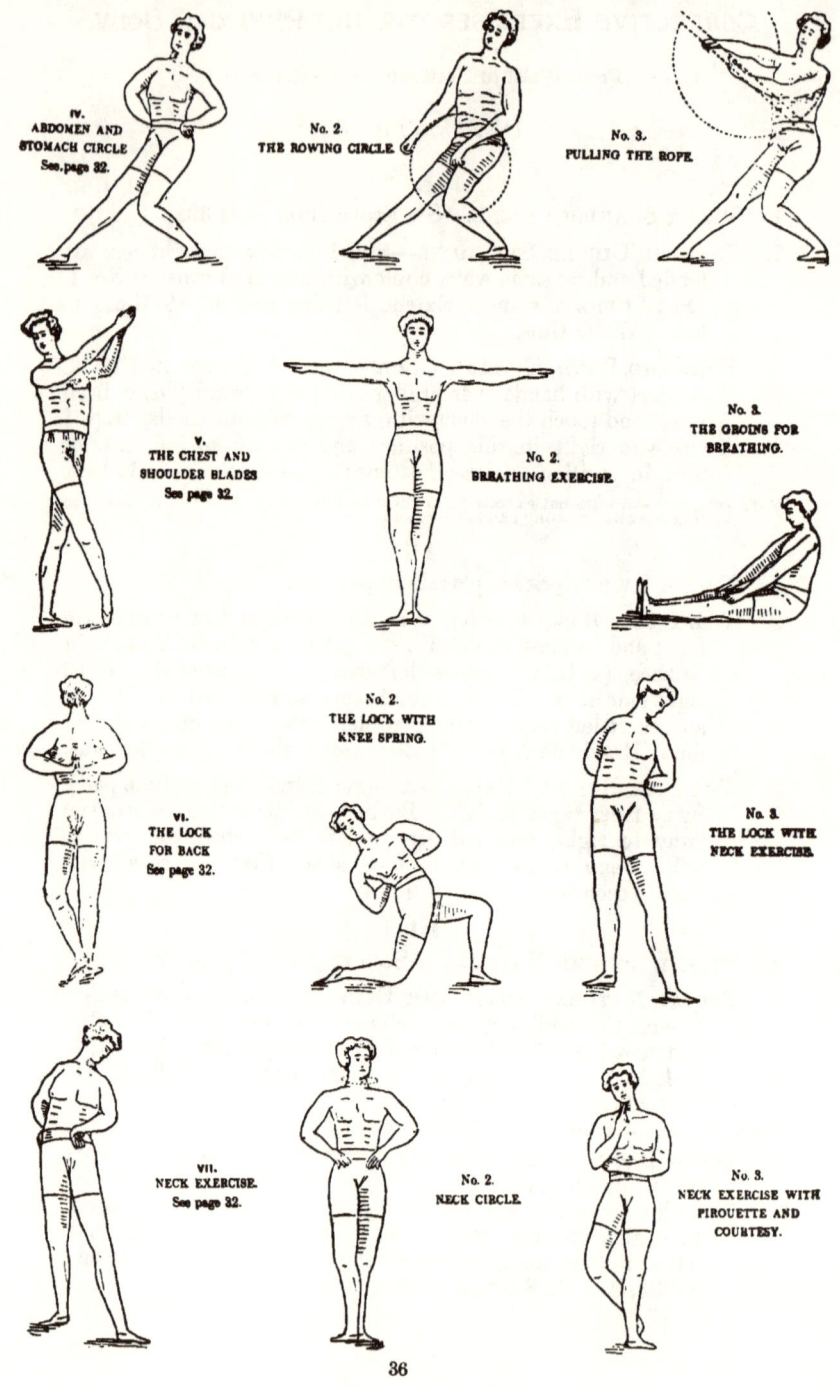

IV.

1. ABDOMEN AND STOMACH CIRCLE.—See explanation page 32.
2. ROWING CIRCLE.—Do No. 1 with this addition, use hands as in rowing. The hands closed, starting at waist, move them downward, next outward, then up to waist. Do 4 times right and left. Repeat 4 times. 2 or 4 bars to each row. Waltz time.
3. PULLING THE ROPE.—Do No. 1, with this addition, use hands as in pulling a rope. The hands closed resting at waist, (body sinking,) now reach up, make a circle of hands (as per trace line) open hands, then close and pull downward diagonally to left side of waist, (if right foot is extended.) Repeat 4 times, now change foot and pull to right side of waist 4 times. 2 bars to reach up, 1 bar down and 1 to recover. Waltz time.

V.

1. THE CHEST AND SHOULDER BLADES.—See explanation page 32.
2. BREATHING EXERCISE.—Inhale as you slowly raise arms, and hold. Now slowly exhale as the arms return to place, make a sucking and blowing sound as you inhale and exhale. Repeat 4 times. Gavotte music. See paragraph 5, page 31, this division.
3. THE GROINS FOR BREATHING.—Lie flat upon the floor, hands at side, rise to a sitting posture without assistance of the arms or hands, and return to horizontal position. Gavotte music, 4 counts to rise and 4 counts to return. Do 8 times.

ADDITIONAL.—From sitting posture, cross the feet and rise to a standing position, without assistance of the arms or hands, and return.

VI.

1. THE "LOCK" FOR BACK.—See explanation page 32.
2. THE LOCK WITH KNEE SPRING.—Lock hands at back and do Knee Spring (IV. page 20,) which is, spring and kneel as in cut, to right. Front view, kneeling sideways. Bend backward when kneeling. Repeat to left. Do alternate 4 to 8 times. Count 2 bars polka music to each kneel.
3. THE LOCK WITH NECK EXERCISE.—Lock hands at back and do Neck Exercise (which follows, and is No. 1, of VII.)

VII.

1. NECK EXERCISE.—See explanation page 32.
2. NECK CIRCLE.—Head upright and do not disturb shoulder muscles, now circle the neck as per trace line. The head should move forward and go around the circle. Relaxation should be perfect. Repeat to left. Do 4 times. Gavotte music.
3. NECK EXERCISE WITH PIROUTTE AND COURTESY.—No. 1 Neck Exercise to right and left, count 1, 2, 3,—1, 2, 3. Then Pirouette right, count 1, 2, 3. Then assume attitude as per cut, (sink down and up,) count 1, 2, 3. In cut, one hand is at chin and the other at elbow. The head leans sharply to right or left. Repeat 8 times. Polka time.

SEVEN LAWS FOR THE PHYSICAL CORRECTIONS OF THE BODY.

(Waist to Neck inclusive.—Series II.)

THE WAIST.

Remember that the "Control of the centre (waist) and freedom of the extremities" will cultivate Reserve Power, or storage battery, as it were, from which to draw when the natural supply is overtaxed, and you should be chary of drawing on this storage battery.

THE TRUNK.

The first thought of the mind in formulating a poise should be Gravity and protection of the Trunk. The muscles under the arms should be developed, as by their strength the shoulders are strengthened to give the arm the power to protect the Trunk or body. The Trunk should be protected because therein is the engine of life, the heart. The arms are its natural protectors from injury.

THE SPINAL COLUMN.

The Spinal Column should never be taxed beyond its natural scope on account of its peculiar construction. It is the only combination of muscle and bone, and the many bones can easily be strained by counter wedging themselves, whereupon something must give way.

THE ABDOMEN AND STOMACH.

The exercises given are for curved outline and subjective control. Delsarte says: "Elevate the spiritual and subdue the physical." This does not imply to kill the physical, not even to mortify the flesh as would the fanatical Brahman, but the spiritual should dominate the physical. The spiritual element can hardly manifest much force through an emaciated or weak physique.

The Chest.

The Carriage of the Chest is the most important (next to the head) for beauty and perfection of the carriage of the body. The Chest assumes its proper position first, the head finishing the gravity of the body. Observe the Law of Opposition of chest and chin, that is, "chest out and chin drawn in;" the next is the harmony of the shoulder blades, thrown back or balanced. The three harmonies that constitute the correct carriage of the Chest, and should be followed, are the chest, the chin and the shoulder blades.

The Back.

The Back should be perfectly straight and smooth, that is, the shoulder blades be almost able to meet and should be strengthened to give the Spinal Column strength. This is accomplished by using the "Lock" Exercise whenever opportunity offers. This exercise strengthens the back bone, upon which depends the chest for its support. Without a strong back bone, round shoulders are sure to develope, and should be avoided.

The Neck.

The Neck muscles under the ear should be strengthened by using the "Neck Exercise." The vertebra at the back of neck, bears the same relation to the head, as the spinal column does to the trunk. The vertebra is assisted (in holding the head in its proper poise) by developing such muscles as will hold the head back in its place, namely : the head should be aplomb on the point of the Spinal Column. The Neck is a connecting hinge, as the wrist to the hand and the ankle to the foot. Flexibility at these joints should be perfect.

Corrective Exercises of the Physical Body.

Series III.

Arms, Shoulders, Elbows, Wrists, Hands, Fingers and Head.

Physical Corrections of the Body.

(Arms, Shoulders, Elbows, Wrists, Hands, Fingers, and Head.—Series III.)

1. Arms.

The Arm complete is the best separate medium of bodily expression. The harmony of curved lines, spiral motion and the line of beauty (wave line) can all be expressed through the arm. Care should be taken to have a perfect harmony of shoulder, elbow and wrist muscles in all arm movements.

2. The Shoulder.

The Shoulder (proper,) relating to the arm, is the seat of the physical power of the arm (see Anatomy of Elements of Man.) From here should generate the power of motion of the arm. As the hips serve to the lower extremities, so do the shoulders relate to the arm in turning of hand outward and all movements of the arm.

3. The Elbow.

Upon the proper training of the elbows depend the beauty and grace of the arm. It contains the seat of the emotional element of the arm (see Anatomy of Elements of Man.) So finely may the harmony be established between the proper bending or blending of wrist and finger tips and shoulders, that the curves of the arm will have no angle whatever at the elbow, which is ordinarily an angle.

4. The Wrist.

The Wrist should receive corrective exercise before that of the hand. It is the machine to which is attached the element known as the intelligence or mentality of the arm, namely, the hand. Its importance is similar to the neck in relation to the head, or ankle to the foot. The exercising should be carefully finished, as any friction here would mar the expression intended by the hand in its great range of movements.

5. The Hand.

The expressive and psychic resources of the hand were well known among the ancient astrologers. This mental instrument of the arm, to the expert palmist " talks." It is substituted for the power of speech and is used by mutes. The student trying the following will easily prove this statement, viz: express with the hand " smoothness," " to scratch," " surprise," " pity," " fear," " determination." The hand almost seems to speak these expressions. It is not only the mental element of the arm, but the outlet (connecting with outer forces) of personal power of expression of the soul, from the fountain head of mentality, namely the brain. Therefore, the hand should be full of intelligent expressions.

6. The Fingers.

The fingers are the eyes of the hand. Their sensitiveness to the slightest touch or change of feeling can be, or should be expressed by their action. You will find each finger seems to have an individuality of its own. It is thus expressed in the " Beauty Position " of the hand. The fingers by their fineness of curve, always finish the movement, or curve of the arm. No curve of the arm is complete without this addition of the fingers.

7. The Head.

If the student has followed with any degree of intelligence to this point, he will understand that here, of all parts of the body, the head is the grandest of all instruments. The *head* of the *body*. Here is the seat of the mind which is the mental control of all other mental elements in the different parts of the body, and they are subject to this master instrument. The head also contains the eyes, which are truly the windows of the soul, the study of which will be found later, and the mouth, which instrument is so necessary to our inner machinery; then the nose, and in fact here is located all the agents of our principal machinery. What is manufactured at their respective factories within, so the product will be shown by these agents. This part of the study is on the physical plane and relates to our physical being. As the head is carried so will we carry the entire body. The head is the heaviest separate instrument of the body. In all our work, its carriage should never be neglected. It cannot help but express something in any position. It is the god-head of our entire body. To impress its importance on the student is the main object of these remarks.

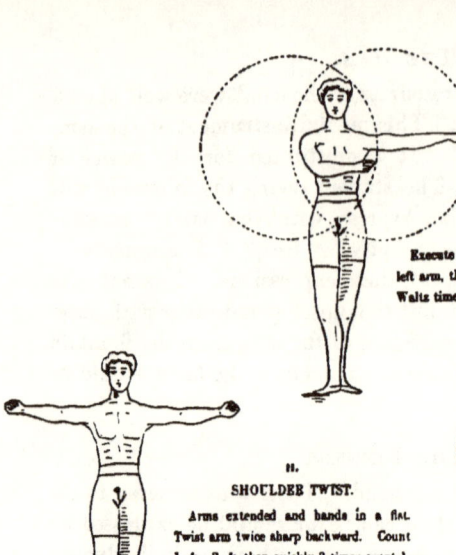

I.
ARM CIRCLE.
Execute a circle with right arm, then left arm, then both arms. 8 times each. Waltz time.

II.
SHOULDER TWIST.
Arms extended and hands in a fist. Twist arm twice sharp backward. Count 1. &—2. &, then quickly 3 times count 1, 2. 3. Do 8 times. Polka time.

III.
ELBOW TWIST.
Assume this attitude, throw the elbows sharp backward and easy forward. Regular rythm 16 times. March time.

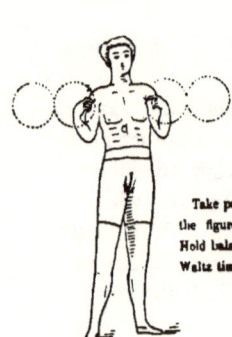

IV.
THE WRIST.
Take positions of wrists thus and trace the figure 8, using wrist to finger tip. Hold balance of arm nearly rigid 16 times. Waltz time.

V.
THE HAND.—(Beauty Position.)
Assume this position (see law) and retain position while executing a wave movement of entire arm. 32 bars waltz music.

VI.
FINGERS.—(Separate Control.)
Fingers all closed, (hand open.) Separate the first finger from each hand, next divide the fingers, next the little finger is alone, now reverse. NOTE.—Do with both hands. Repeat 4 times. Schottische time.

VII.
HEAD.—(Side to Side.)
Take this position and throw sharply to right, then to left, count 1 &. Then quickly to right, left, count 1, 2. 3. Do 4 times. Polka

Seven Corrective Exercises of the Physical Body.

(Arms, Shoulders, Elbows, Wrists, Hands, Fingers, Head.—Series III.)

	TITLES.	CORRECT.	RELATION.	HARMONY.
I.	Arm Circle.	For Ease of Motion. Strength of Shoulders.	To Shoulder Blades and the Back. Protection of Trunk.	Established for the Expression of the Mind and Magnetic Channel.
II.	Shoulder Twist.	Turning out of Arm.	To Wrist. Base for Head. Muscles under Arm.	Shoulder Blades, and Wrist. Muscles under Arm and Ebolw.
III.	Elbow Twist.	Angle at Elbow.	Shoulders and Wrist must be assisted by Elbow.	Emotion of Arm, Harmony of Fingers and Shoulder Blades.
IV.	The Wrist.	Flexibility.	Shoulder Muscles and use of Hand.	Shoulder to Elbow, to Wrist always.
V.	The Hand Beauty Position.	Flexibility. Intelligence.	Expression of Arm and Mind.	Physical Expression.
VI.	The Fingers Separate Control.	Individuality of Fingers.	Beauty and Intelligence and outlet of Power.	As Eyes to the Mind. Mental seat of the hand. The finish of the curves of arm.
VII.	Head.	Ease of Poise.	Vertebra and Chest.	The Fountain of Expression.

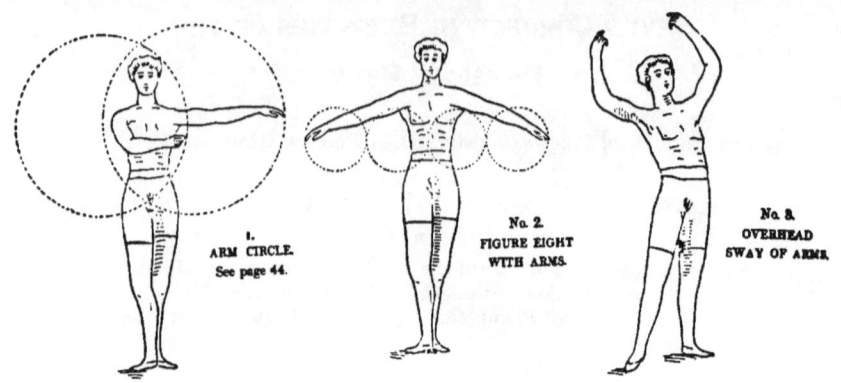
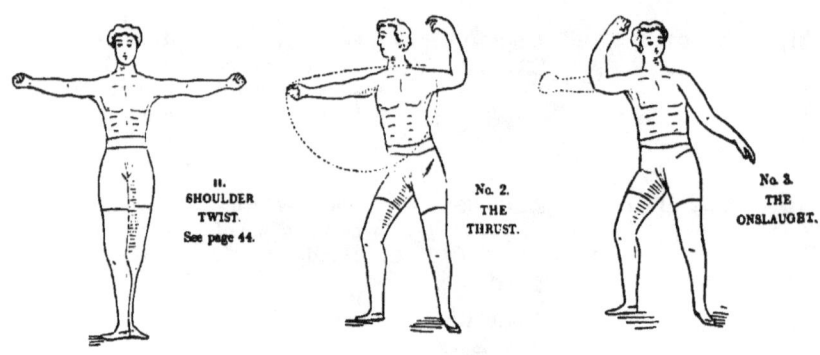

Corrective Exercises for the Physical Body.

(Shoulders, Arms, Elbows, Wrists, Hands, Fingers and Head.—Series III.)

DERIVATIONS.

I.

1. ARM CIRCLE.—See explanation page 44.
2. FIGURE EIGHT WITH ARMS.—Execute the figure eight, horizontal, per trace line using entire arm. Do same perpendicular. Do both 8 times. Waltz time.
3. OVERHEAD SWAY OF ARMS.—Place arms overhead and sway from right to left side, bending waist. Observe beauty position of hand. Do 16 times. Repeat same and walk forward and backward. 2 bars of music for each sway. Waltz time.

II.

1. SHOULDER TWIST.—See explanation page 44.
2. THE THRUST.—Take attitude as per cut, do arm circle per trace line and *thrust*. At thrust of right arm stamp right foot. The thrust should be an out twist of the entire arm, the other arm is up. Repeat left arm and left foot. Do 16 times each side. Count 1, 2, 3, for each thrust. Polka time.
3. THE ONSLAUGHT.—Same attitude and movement as the thrust, only let arm execute a complete circle and stop suddenly in attitude as per cut, stamp with foot. The same music, etc. Hand clinched.

III.

1. ELBOW TWIST.—See explanation page 44.
2. ELBOW TWIST WITH ANKLE EXERCISE.—Do Elbow Twist, at same time doing Ankle Exercise. Ankle Exercise is No. 2, Series I. page 20. Do 32 times. March time.
3. ELBOW TWIST WITH BUOYANCY EXERCISE.—Do Elbow Twist at same time doing Buoyancy Exercise. Buoyancy Exercise is No. 5, Gravity, page 10. Do 32 times. March time.

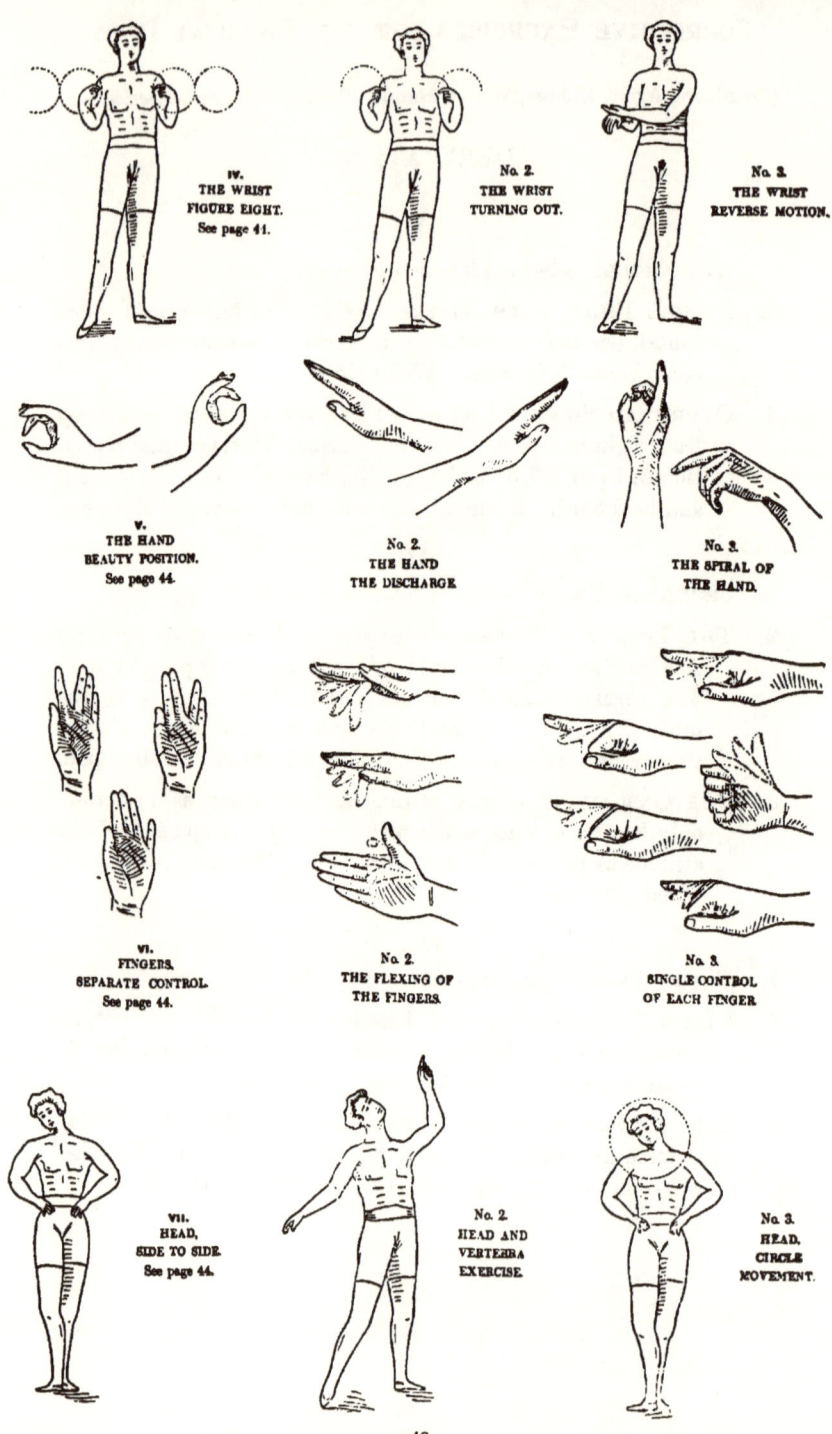

IV.

1. THE WRIST.—See explanation page 44.
2. THE WRIST, TURNING OUT.—Assume hand and wrist position per cut, and turn them sharply out (opening hand) do twice, count 1 & 2 &, then quickly 3 times, count 1, 2, 3. The force of the hand is generated at wrist. Do 16 times. Polka time.
3. THE WRIST—REVERSE MOTION.—Place back of hands together as per cut, now move fingers out and toward you, then have all fingers point downward (always keep back of hands together.) Now have original top hand move outward and other hand inward, until the hand originally on the bottom is on top, as per cut. Repeat from this position. Do continuous 16 times. For each complete motion, count 1, 2, 3. Polka time.

V.

1. THE HAND BEAUTY POSITION.—See explanation page 44.
2. THE HAND DISCHARGE.—Close hands in a fist and open them suddenly with force, see cut, do 2 times, count 1 & 2 &, then 3 times quickly, count 1, 2, 3. The force should be generated from the shoulders to wrist. Do 16 times. Polka time.
3. THE SPIRAL OF THE HAND.—Assume drooping position, per cut. Execute a spiral movement of hand, the index finger leading and the force starting from the wrist. The hand in this spiral movement *upward* takes the action of putting on a glove. (Base of thumb being drawn to base of small finger.) Observe Beauty Position. Recommence by relaxing to starting point. Balance of arm partly rigid. 32 bars waltz rhythm.

VI.

1. THE FINGERS—SEPARATE CONTROL.—See explanation page 44.
2. THE FLEXING OF FINGERS.—Hold hands up and crook each finger and thumb repeatedly, by itself, (flexing it.) Do same finger of each hand at same time, then flex all fingers at once. Dramatic Tremulo music.
3. THE SINGLE CONTROL OF FINGERS.—Close hands and simply extend one finger at a time and draw it back, (flexing it,) extend same finger of each hand at same time. Do 4 times. Schottische time.

VII.

1. HEAD, SIDE TO SIDE.—See explanation page 44.
2. HEAD AND VERTEBRA EXERCISE.—Spring to right and assume attitude, per cut. The head thrown sharply back, looking up. Move arms from 1st to 2d arm position to attitude. Repeat to left. Count 2 bars to each side, when proficient 1 bar. Do alternate 8 times. Polka time.
3. HEAD CIRCLE.—Drop the head in front, circle to right per trace line. At back it should hang as far as it did in front in executing the circle. Repeat to left. Note.—Eyes should not move, and shoulders and trunk should be rigid. Do this 4 times each way. No music.

SEVEN LAWS FOR THE PHYSICAL CORRECTIONS OF THE BODY.

(Arms, Shoulders, Wrists, Hands, Fingers, Head.—Series III.)

THE ARMS.

The Arms should have freedom of motion. They should never rest against the trunk; or rather all arm motions should be from the shoulders, which require the strengthening of the Torso or muscles under the arm.

THE SHOULDERS.

The Shoulders should always be let down and "set" until a feeling of comfort or adjustment is felt throughout the body in any poise.

THE ELBOWS.

A perfect harmony must be cultivated with wrist, fingers and shoulders, and not until then will the angle of the elbow be lost.

THE WRIST.

The Wrist should be flexed before the hand, and must be applied to all individual motions of the hand to allow it the expression intended. The Wrist should lead the movements of the hand, not forgetting the harmony of the other parts of the arm.

THE HAND.

The Hand should always give the intention and expression of the movement. It is the mind of the arm. The latitude or range of the hand is only second to the mind, in expressing physical motion.

THE FINGERS.

The fingers express the fineness or "the art of" the expression of the hand, as the eyes do of the mind. Finish the curves of the arm with the fingers. Personal magnetism from the body to the mind is mostly distributed through the finger tips. The thumb should never be drawn into the palm of the hand, as that denotes weakness; rather curve the thumb, creating a harmony of both joints. In executing the curves of the arm, the fingers trace the outer line of the circle, and at the horizontal line only do they straighten out.

THE HEAD.

The Head is the law-maker of our body. No known law governs its action. It contains the essence of all our elements and all our machinery. The ideas formulated there, register their expression by the poise of the Head. It is the fountain or source of all intention, here they are received and sent out. Aside from its intellectual and life elements the Head should set well back The chin drawn in (or chin in opposition to chest; law is chin in, chest out.) As the head is carried so will you carry your whole body, and express all movements of the body. The Head is the register of that within, and therefore the mind must govern all. The Head is expressive in any position.

The Anatomy of the Elements of Man.

The Trinity.

THE ANATOMY OF THE ELEMENTS OF MAN.

One of the basic principles of the Christian religion is the Trinity of the Creator.

GOD
THE FATHER, SON AND HOLY GHOST.

THE TRINITY
Consists of Three Forces

LOVE,   WISDOM,   POWER,

Which is The Grace of God.

COUNTERPART.

MAN

THE SOUL,   THE MIND,   THE BODY.

The Grace of Man

Consists of Three Elements

EMOTIONAL.   MENTAL.   PHYSICAL.
(or Moral.)   (or Reason.)   (or Vital.)

THE UNIVERSE.

AIR,   WATER,   EARTH.

GOD, MAN AND THE UNIVERSE,
THREE IN ONE

GOD IN ALL.

THE SOUL, THE PHYSICAL BODY AND GRACE.

THE TRINITY OF THE SOUL,
is

The INTELLIGENCE force.
The LIFE FORCE within us.
The SPIRITUAL power.

This Trinity in One is the Soul—

DUALITY.

INTELLIGENCE is WISDOM. ⎫
LIFE FORCE is POWER. ⎬ See Trinity of God.
SPIRITUAL FORCE is LOVE. ⎭

Thus do we resemble the Great Creator only on a lesser scale.

THE PHYSICAL BODY,
embraces

GRAVITY, which relates to the Mental.
FLEXIBILITY, which relates to the Emotional.
FORCE, which relates to the Physical.

GRACE.
is

THE EXPRESSION OF THE SOUL,
THROUGH THE MIND,
BY MOTION OF THE BODY.

TRINITY.

Notwithstanding our advanced civilization we are constantly confronted with the old philosophic question as to " that invariable existence of which these are variable states? " Concerning this there have been sundry hypotheses. Mine being primarily that everything, material and otherwise, is based upon the Trinity. I find a trinity in man and in the most minute details, into which he may be sub-divided; the same holds true in nature.

The Trinity is the rock on which this work is built, and the analogy to all the laws, works and movements herein are clearly shown.

Man consists of the Soul, the Mind and the Body.
REFLECTION.
The Mind is of the Father, The Body is of the Son,
The Soul is of the Holy Ghost.

THE SOUL.

The mention of Duality, (page 55), is to show its reflection and " *part of* " the all prevailing
" GOD. "

As a sunbeam from the heavens above coming millions of miles to earth is heat or " power " so is the Reflection of the Grace of God to Man.

" POWER IS NEVER LOST,"
as it continues in some form or another <i>ad infinitum.</i></p>

WATER TO STEAM.
STEAM TO AIR.
AIR TO WATER.

THE ANATOMY OF THE ELEMENTS OF MAN.

Their Location in the Physical Body.

MAIN FORCES.

THE MENTAL.   THE EMOTIONAL.   THE PHYSICAL.

LOCATION.

I. THE MENTAL—is in the HEAD—the Brain.

II. THE EMOTIONAL—is in the TRUNK—the Heart.

III. THE PHYSICAL—is in the LIMBS AND ARMS—propelling power.

DERIVATIONS.

THE HEAD.

IV. MENTAL—the Forehead—Thought.
EMOTIONAL—the Eyes and Nose—Expression.
PHYSICAL—the Lower Jaw—Action and Strength.

THE TRUNK.

V. MENTAL—the Lungs—Breathing.
EMOTIONAL—the Heart—Seat of Emotion.
PHYSICAL—the Abdomen—Animal.

THE LEGS.

VI. MENTAL—the Foot—expressing the propelling force.
EMOTIONAL—the Knee—Bouyancy.
PHYSICAL—the Hip—The power to propel.

THE ARM.

VII. MENTAL—the Hand—Meaning; touch.
EMOTIONAL—The Elbow—Gives expression which is emotion.
PHYSICAL—the Shoulders—power and strength.

The Anatomy of the Elements of Man.

Sub-division—The Head.

The Forehead.

Mental—the Brow—Knowledge.
Emotional—the Eyebrow—Expression.
Physical—the Hair—Protection.

The Eye.

Mental—the Pupil—Sight.
Emotional—the Iris—Expression.
Physical—the Eyelid—Protection.

The Mouth.

Mental—the Tongue—Taste.
Emotional—the Lips—Feeling.
Physical—the Jaw and Teeth—Masticate.

The Nose.

Mental—the Point—Indication.
Emotional—Membranes—Sensitive.
Physical—the Bridge—Strength.

The Ear.

Mental—Auditory Nerve—Transmitter.
Emotional—Ear Drum—Receiver.
Physical—Casement—Protection.

The Throat.

Mental—the Palate—Mental Taste.
Emotional—the Larynx—Voice.
Physical—Windpipe—Air Passages.

A Hair.

Mental—the Point.
Emotional—its Vesicle.
Physical—the Root.

The Anatomy of the Elements of Man.

Sub-division—The Trunk.

The Heart.

Mental—Auricles and Ventricles—receives the Blood.
Emotional—Endocardium—Membrane forming inner Valves.
Physical—Pericardium—Sac containing Heart and Root of the Great Vessels.

The Lungs.

Mental—the Aëration Surfaces.
Emotional—Air Cells.
Physical—Root, (Pulmonary Artery and Nerves.)

The Abdomen.

Mental—the Pelvic Organs.
Emotional—the Glands and Ligaments.
Physical—the Rectum.

The Back.

Mental—the Spinal Column.
Emotional—the Shoulders.
Physical—the Back Bone.

The Spinal Column.

Mental—the Point.
Emotional—the Connecting Cords.
Physical—the Base.

The Neck.

Mental—Front—Sensitive.
Emotional—Sides—Curve.
Physical—Back of Neck—Base.

The Stomach.

Mental—Mucus Lining.
Emotional—Peptic Glands.
Physical—Intestines.

The Anatomy of the Elements of Man.

Sub-division—The Extremities.

THE LEG.

The Foot.
MENTAL—the Toes.
EMOTIONAL—the Instep.
PHYSICAL—the Heel.

The Knees.
MENTAL—the Joint.
EMOTIONAL—the Ligaments.
PHYSICAL—the Knee Cap.

The Hip.
MENTAL—the Ligaments.
EMOTIONAL—the Socket.
PHYSICAL—the Haunch Bone.

The Toes.
MENTAL—the Big Toe.
EMOTIONAL—the Little Toe.
PHYSICAL—the Middle Toes.

The Heel.
MENTAL—the Calcis—point.
EMOTIONAL—the Ligaments.
PHYSICAL—Synovial Protector.

The Ankle.
MENTAL—the Joint.
EMOTIONAL—the Ligaments.
PHYSICAL—the Shin Bone.

The Thigh.
MENTAL—the Point (at Knee.)
EMOTIONAL—the Femur.
PHYSICAL—the Socket-ball.

THE ARM.

The Hand.
MENTAL—the Fingers—Touch.
EMOTIONAL—the Palm—Greeting.
PHYSICAL—the Back—Striking.

The Elbow.
MENTAL—the Point.
EMOTIONAL—the Joint.
PHYSICAL—the Ligaments.

The Shoulder.
MENTAL—the Joint.
EMOTIONAL—the Ligaments.
PHYSICAL—the Shoulder Blades.

The Wrist.
MENTAL—the Joint.
EMOTIONAL—the Ligaments–Pulse.
PHYSICAL—the Band of Wrist.

The Fingers.
MENTAL—the Index Finger.
EMOTIONAL—the Little Finger.
PHYSICAL—Middle Fingers and Thumb.

A Finger.
MENTAL—the Tip.
EMOTIONAL—the Middle Joint.
PHYSICAL—the Knuckle.

A Finger Nail.
MENTAL—the Cuticle.
EMOTIONAL—the Nail (proper.)
PHYSICAL—the Hoof (protection.)

The Location of the Elements of Man.

The Main Forces.

In the location of the elements in the physical body, attention is called to the main forces: namely to the Head—the mental, to the Trunk—the emotional, and to the Extremities—the physical.

Next, to the sub-division of each separate part. The relation and harmony is so complete as to be a "*part of all.*"

The Head.

The Head is the seat of Mentality. All its functions are but the mental agents of their respective instruments or machinery: *i. e.* the mouth for the stomach, the nose for the lungs, the ears for the mind, (brain,) and the eyes for the heart.

The Trunk.

The Trunk is the seat of Emotion. All instruments herein are related to, or create emotion, *i. e.* the lungs for breathing, the heart for feeling, and the abdomen for its emotions.

The Extremities.

The Extremities are the seat of Physical Power. The Arms are for protection, and the Limbs are for locomotion.

Attention is also called to the fact that the grosser elements are located below the heart, and the finer are above this centre. The " heart " is the source from which radiates our life force around it.

The Sun is the Heart of the Solar System.

The Study of Curved Lines.

Circles, Wave Lines and Spiral Movements.

The Study of Curved Lines.

I.

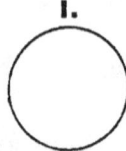

CIRCLES. Mental. THE MIND.

The Circles are the mental movements. The mind should be inoculated, or thought should be concentrated, to express two intentions — Circles and Grace. These thoughts are transmitted to the finger tips and hand.

These exercises are subjects of intelligence. The student knowing this, will be able to transmit corresponding action to the body movements, also receive expression from the heart and thus impart these elements to the movements.

II.

WAVE LINES. Emotional. THE BODY.

The Wave Lines are the emotion of movements. Here the mental thought becomes intensified, and the feeling of the heart is included. This emotion may properly be called the reflection of the soul, as it comes from the life source, namely the heart.

In Movements of Curves it is necessary to take the student back to "The Corrective Exercises of the Physical Body," which should be well mastered. These beautiful movements will then show the reflection (grosser to finer) of their training and perfection. They now have the assistance of the Soul and Mind.

III.

THE SPIRAL MOTION. Power. THE SOUL RESPONDS.

The Spiral Motion is the Power necessary to complete this Trinity. The Spiral Motion here meaning Spiritual Force. The Spiral Motion when added to these curves and movements, gives strength, also feeling of heart, which opens the channel through which flows our Personal Magnetism, and the Soul responds.

The flow of Personal Magnetism of the Soul through these Arm and Body movements, will in a short time pervade our whole being. The student will feel the Soul presence; it will reach your auditor and he will feel it. We are then adding to ourselves "The Grace of Man."

CURVED LINES.

Consist of

1. CIRCLES 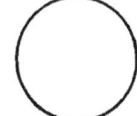 Mental—Gravity.

2. WAVE LINES  Emotional—Flexibility.

3. SPIRAL MOVEMENT  Physical—Force.

DERIVATIONS.

4. CIRCLES with Straight Attitudes.

5. CIRCLES with Arch Attitudes.

6. WAVE LINES with Circle Lines and Spiral Movement.

7. SPIRAL MOVEMENTS with Knee Bend Attitudes.

The Alphabet.

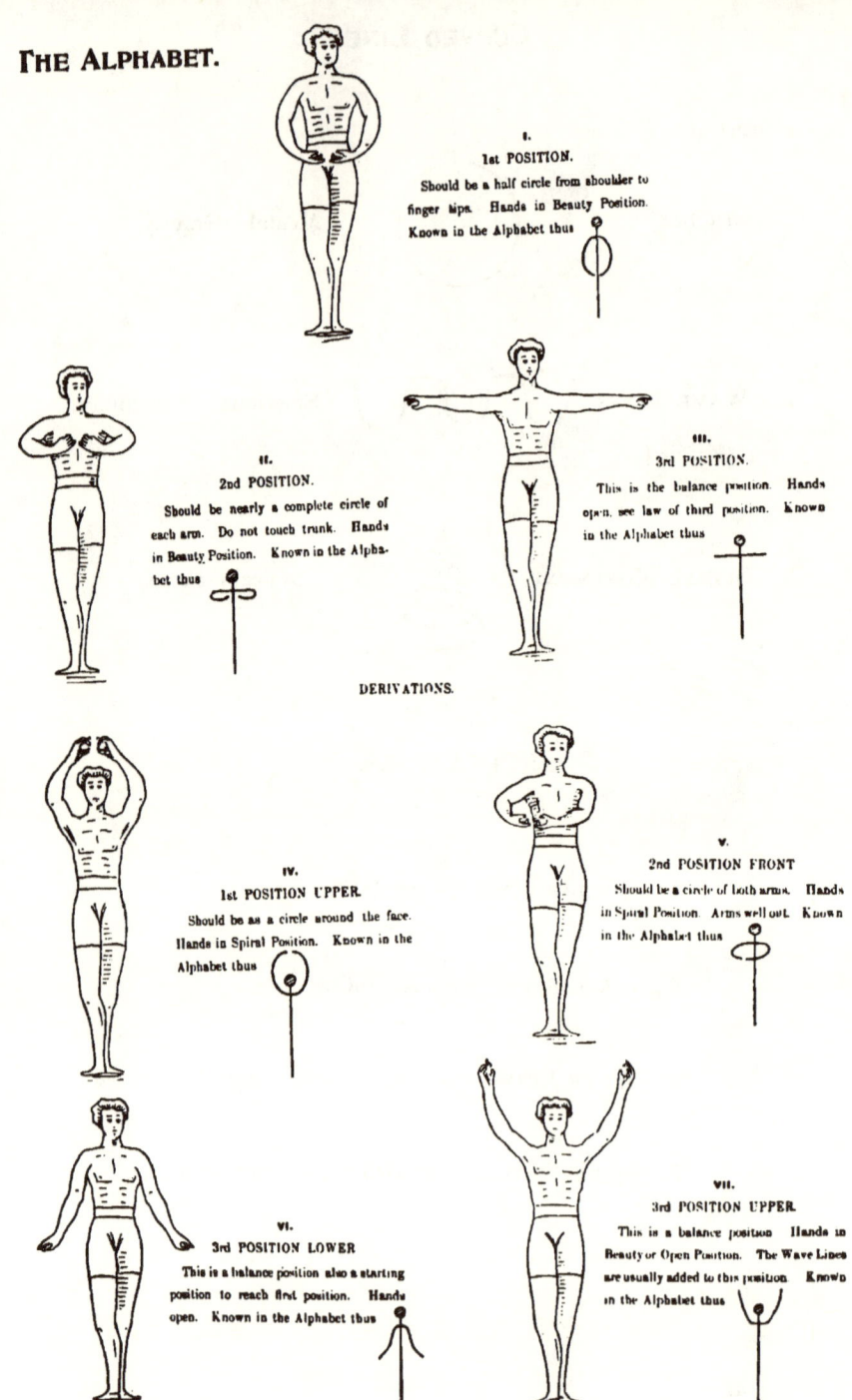

I.
1st POSITION.
Should be a half circle from shoulder to finger tips. Hands in Beauty Position. Known in the Alphabet thus

II.
2nd POSITION.
Should be nearly a complete circle of each arm. Do not touch trunk. Hands in Beauty Position. Known in the Alphabet thus

III.
3rd POSITION.
This is the balance position. Hands open, see law of third position. Known in the Alphabet thus

DERIVATIONS.

IV.
1st POSITION UPPER.
Should be as a circle around the face. Hands in Spiral Position. Known in the Alphabet thus

V.
2nd POSITION FRONT.
Should be a circle of both arms. Hands in Spiral Position. Arms well out. Known in the Alphabet thus

VI.
3rd POSITION LOWER.
This is a balance position also a starting position to reach first position. Hands open. Known in the Alphabet thus

VII.
3rd POSITION UPPER.
This is a balance position. Hands in Beauty or Open Position. The Wave Lines are usually added to this position. Known in the Alphabet thus

The Study of Circles, Wave Lines and Spiral Movements.

The Seven Arm Positions.

Alphabet.

I. 1st Position. Base position. (Physical.)

II. 2nd Position. Circle position. (Emotional.)

III. 3rd Position. Horizontal position. (Mental.)

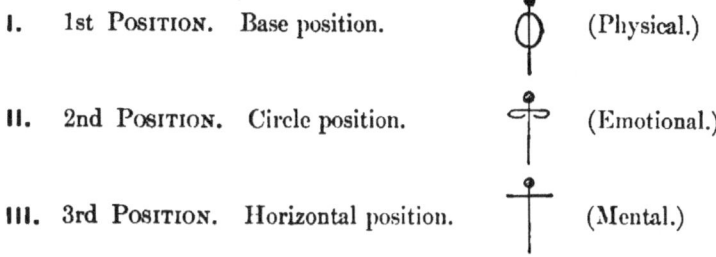

Derivations.

IV. 1st Position Upper.

V. 2nd Position Front.

VI. 3rd Position Lower.

VII. 3rd Position Upper.

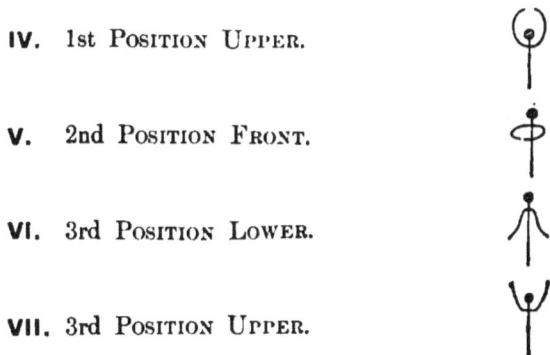

PRACTICE.

The three positions of the Arms should be thoroughly mastered, after which the four Derivations may be easily memorized, as their elements are all embraced in the three positions. No music.

These Arm Positions will be found to cover all Arm Movements— in Circular Lines, Wave Lines, Spiral Movements and Attitudes.

CIRCLES 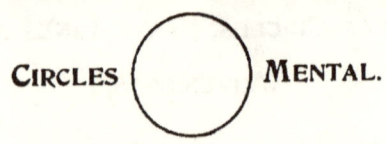 MENTAL.

1. THE HARMONIC THREE.

All exercises must be a slow continuous movement. The Mental Expression in this exercise is Grace, and should be transmitted to and through the fingers (the mental of the arm.) The head should impress the expression of Grace in all movements of Grace.

2. THE COMPLETE CIRCLE.

This exercise is of the same order as No. 1, only making a complete circle of each arm. The mental expressions are Grace and Circles, and are transmitted the same way,—Grace by expression, and Circles by the tracing of an invisible circle with the finger tips, (see Laws, this series.) The eyes follow the hands up, when they pass above the line of vision.

3. THE ALTERNATE EXERCISE. (Harmony of 3rd position.)

This exercise consists of executing No. 2, with one hand, while executing No. 1 with the other hand, and repeating it. Then showing the 3rd or Balance position, which establishes the blending of continuous movements, (see Laws of this series.) The mental expression is Grace and Circles. When Attitudes are added to these arm movements they are illustrated, and should be accented by holding them.

4. THE CIRCLES OF GRACE.

The Circles in this exercise move both ways for each arm; and are known as the Circles of Grace. The mental expression is a continuous thought of Grace and Circles, no break occurring until you cease executing the same. The movement of the hands assumes the Wave Lines, which add emotion and therefore intensifies the expression of Grace.

5. The Transmission of Power.

In this exercise the Mind is given a Force to Express. These thoughts are accented at certain parts of the movements. In this instance "*To gather Power*" is at 1st Upper Arm Position and is held longer. This attitude is illustrated. The Head here reaches a climax expression. The accent again occurs at 2d Arm Position Front, which is illustrated, and is intended to express "*Transmitting the Power.*" In this poise the Head reaches its final climax of intensity of expression.

6. The Repentant.

This exercise not only gives the mind occupancy, but takes in the whole body, and is an impersonation of a *Penitent*. Following the Arm Movements to the first attitude, (illustrated and should be accented,)—is an attitude of sorrow—"*The Penitent,*" which is expressed by the whole body. Blending the arms through the next Arm Positions to the second attitude, (illustrated and is accented,) denotes "*The Cry of Anguish.*" The arms and body dissolving to the last Arm Position, meaning "*hope gone.*" Care should be taken to use Balance Position of Arms at each change, and observe the Laws of Poise.

7. Inspiration, All my Soul I am Content.

This exercise is similar in nature to that of the Repentant, except that it expresses the glorifying kind of Grace. To express it properly the soul feeling must reach that intensity, that you can give it expression. The Soul expression is—"*I give my Soul.*" The mental expression should assist the flow of feeling and direct the intelligence force. The Harmony should be purity of thought, elevation of feeling and control of Force or Reserve Power. The Attitudes are illustrated and should be accented, and may be assumed with a Spiral Body Motion.

SEVEN EXERCISES CIRCLES  MENTAL.

EXERCISE NO. 1.   THE HARMONIC THREE.
Arm Positions.   1st to 2d   to 3d   to 1st   Do continuous.

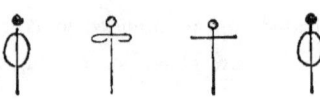

EXERCISE NO. 2.   THE COMPLETE CIRCLE.
Arm Positions. 1st to 2d to 1st up. to 3d to 1st Do continuous.

EXERCISE NO. 3.

THE ALTERNATE EXERCISE.   (Harmony of 3d Position.)
Arm Positions.—Use both arms at same time.
Right 1st to 2d to 1st up   to 3d to 1st to 2d   to 3d to 1st
Left 3d l. to 1st to 2d.   to 2d to * 1st up.   to 3d to 1st

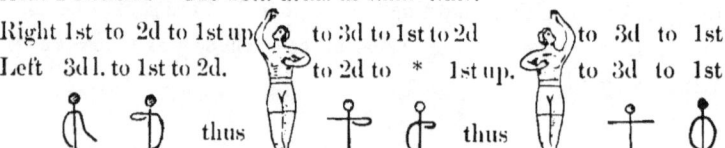

EXERCISE NO. 4.   THE CIRCLE OF GRACE.
Arm Positions.

   1st to 2d   1st up.   to 3d   to 1st   to 2d   to 1st up.   to 2d

   to 1st   to 3d   to 1st up.   to 2d   to 1st   to 3d l.

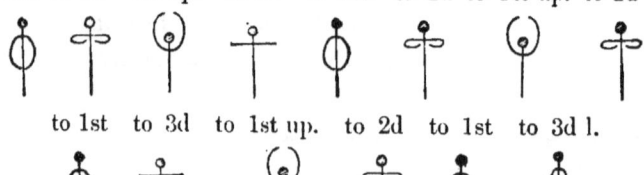

* Hold 2d position.   up. upper.   l. lower.

EXPLANATION.

1. THE HARMONIC THREE.

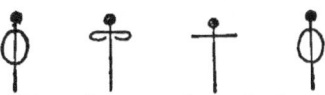

Standing poise. Move the arms through the three positions as per cut, making a slow continuous movement. Hands are in Beauty Position and open at 3d Position. The hand turns (palm down) at 3d Position (see Laws this series.) Count 8 for this exercise once through. Do 8 times. Slow gavotte music.

2. THE COMPLETE CIRCLE.

Standing poise. Move the arms through positions as per cut. As the arms pass from 2d to 1st Upper Position, (before the face,) the eyes follow them up. Same for hands as No. 1. Count 8 for once through. Do 8 times. Slow gavotte music.

3. THE ALTERNATE EXERCISE.

This is Exercise No. 2 with right hand and No. 1 with left hand and reverse, and a Balance Position before the finish. Standing poise. Move both arms through position as per cut, accent by holding illustrated movement. Same for hands as No. 1. The eyes follow the high hand, and the head is away from the highest hand, (see Laws this series.) Observe Circles of each arm, except at Balance Position. Count to accented poise 4, hold accent 4—same count, doing exercise continuous. 32 bars slow gavotte music.

4. THE CIRCLES OF GRACE.

This is No. 2—The Complete Circle—going around and reverse, and should be done continuous, the last position (3d Lower) being a Balance Position to allow starting over again—by blending one position into the other.—The hands at 1st Upper take the Wave Line motion to change the Circles to reverse. The Laws of Head and Arms should be observed, and grace of movement have full play. Count 8 for each complete Circle. 64 bars gavotte music.

EXERCISE No. 5.   THE TRANSMISSION OF POWER.
   NOTE—Accent Attitudes (illustrated) by holding them.

Arm Positions.
1st to 2d to 1st fr. up.   to 2d to 2d fr.   to 3d l. to 1st

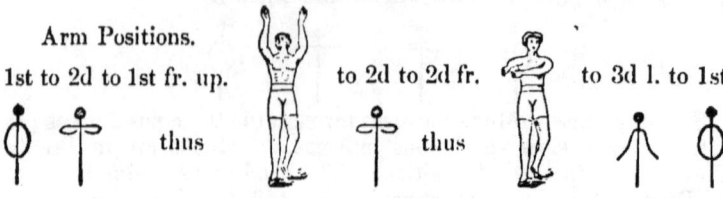

thus   thus

EXERCISE No. 6.   THE REPENTANT.
   NOTE—Accent Attitudes by holding them.
Arm Positions, (both at same time.)

3d l. to 1st to 2d to   l. head   3d to 1st to 2d to 1st f. up   2d to 3d l.
                        r. 3d l.

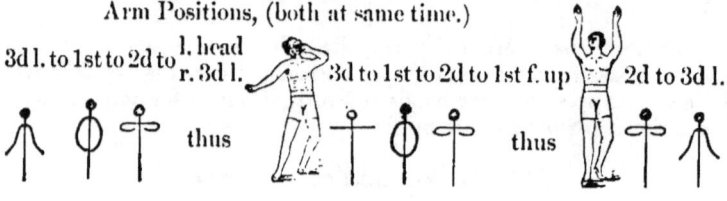

thus   thus

EXERCISE No. 7.
   Subjects—THE INSPIRATION, ALL MY SOUL, I AM CONTENT.
      NOTE—Accent Attitudes.

THE INSPIRATION.   ALL MY SOUL
1st to 2d to 3d up.   to 2d to 1st to 2d to 3d

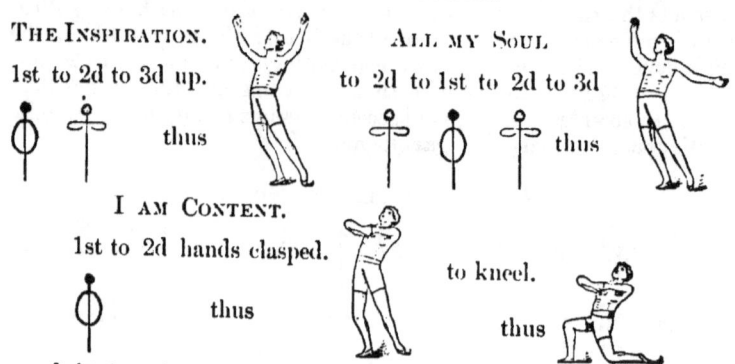

thus   thus

I AM CONTENT.
1st to 2d hands clasped.

thus   to kneel.
        thus

fr. front.   l. left hand.   r. right hand.

72

5. THE TRANSMISSION OF POWER.

Start this exercise from a slight sinking movement. Move arms through positions as per cut to illustration, (accented.) The climax of gathering force is reached in this Attitude, by a mental feeling of reaching upward. This is an Arch (outward) Attitude. Now move the arms through the next position (body natural) to illustration (accented) and here transmit or impart the Soul feeling. Hands assume wave lines at climaxes. All movements should be very slow and impressive. Use an appropriate hymn and move to its rythm. Do 4 times.

6. THE REPENTANT.

Standing Poise. Move arms as per cut to first impersonation, "*The Penitent*" (illustrated and accented.) This is an Arch (inward knee bend) Attitude (Study of Attitudes.) The arms go to head, and 3d lower with the entire body and mentality expressing the sense of the subject. Next, (body natural) the arms pass through the positions to accented Attitude (illustrated) "*The Cry of Anguish.*" This is an Arch (outward) Attitude. The physical and mental expressing the sense, by starting from a slight sinking movement and thus emphasizing a stronger effect. Hands same as No. 5. Now moving arms to last position and the drooping of the head and an inward bend of the body well expresses "*hope gone*" if the mind has been so imbued, (not illustrated.) Count to accented poise 4, hold accent 4; same count until finished. Do 4 times. Slow gavotte music.

7. THE INSPIRATION, ALL MY SOUL, I AM CONTENT.

Start this exercise from a slight sinking movement, rising upward with a spiral motion. Move arms through positions as per cut, to the first accented illustration. These are Arch (outward) Attitudes (¾ angle front view.) The mind must be purged of all carnality and the eyes should reflect the feeling of being inspired. The hands and head with the body, should also have a slight wave or spiral motion to reach these accents. From this Attitude move the arms through the next positions, repeat the sinking and upward spiral motion of the body, and impress the 2d Attitude by an action of the head, expressing "*I give my Soul.*" The same action is repeated to 3d Attitude, (hands clasped); from that position slide rear foot backward and slowly kneel, assuming last poise, eyes looking up. About face (¾ front view) and repeat exercise. Count to accented poise 4, hold accent 4; same count until finished. Do 4 times. Slow gavotte music.

WAVE LINES 〜〜〜 EMOTIONAL.

1. WAVE LINE BALANCE.

The Wave Lines are executed usually in horizontal and diagonal lines and resemble waves; when added to Circles they produce the emotion of such lines. This exercise is a horizontal line movement with each arm. The emotion should start at the shoulders and be transmitted along the curves of the arm. The mental expression is strengthened by a more intense feeling of grace imparted to these wave motions.

2. THE HARMONIC THREE WITH WAVE LINES.

This is Exercise No. 1 of Circle series, (page 68.) The Circles are executed with Wave Lines, giving to them grace, and the Wave Lines emotion. The mental expression is emotional grace, if the method heretofore described has been followed.

3. THE WAVE LINES DIAGONAL.

The Wave Lines here follow a diagonal line of direction. Both arms, starting at one side and moving diagonally across the body and upward, require the head and body to take on these Wave Lines. They are strengthened by starting from a slight sinking of body, to its fullest stretch in a diagonal direction. This is really creating an emotional action of the physical body. The mental feeling of grace should be manifest.

4. THE UPWARD ADVANCE. (A PASTORAL.)

Subjects—GATHERING INSPIRATION, RECEIVING INSPIRATION, INSPIRED.

In this exercise the intention of expressing an emotion of intelligence is obvious by taking the sense of each subject. In assuming the first attitude the entire body should feel the emotion of "*Gathering Inspiration.*" The next attitude the mental expression of "*Receiving Inspiration*" is a combination of fear, joy, uncertainty and certainty in one, or suppressed emotion. At the next attitude the mental and emotional flood gates of feeling are opened and the feeling of "*I am Inspired*" permeates the entire being. The last attitude is an offering of "steadfastness" which is a fitting finale to this Pastoral.

5. The Emotion of Love.

Subjects—The Stars Above. To thee my Heart. At thy Feet.

This conception calls for a strain of Ideality, which is developed Personal Magnetism. (See study of Personal Magnetism and Ideality.) Here the purity or essence of emotion is shown, namely, "*Love.*" In the first attitude "*The stars above,*" is expressed by the glorifying feeling of Ideality and the interpreter of this subject must lose "self" and permit his Soul to roam in the realms of "*the stars above.*" The next attitude expressing "*To thee my heart*" is the transmission of said feeling. The soul flight is then again repeated and the attitude of "*At thy feet*" expresses the goal sought and found.

6. In a Garden Fair.

Subjects—He calls to her. She is in a balcony. His presence there. His heart to her, his lips to her. His worship. His perplexity.

This exercise is similar to No. 5 only of a lighter vein, and gives full play to Wave Lines in a diagonal direction for the head, arms and body, with an emotional intent (a meaning.) Description—The scene is in a Garden Fair. An Apollo gliding toward a balcony whereon stands his goddess; sending forth his whole Soul, he attracts her attention. In the exhilaration of bliss he moves on air; a kiss, a heart of joy, ecstasy—seem to spring from his Soul. His perplexity.

7. The Gateway of Heaven—Death.

Subjects—Discovery and Ignorance. Curiosity and Dawning. Horror. Realization. The Truth—Heaven.

The emotion expressed here is entirely of a Soul nature. The physical is subdued or mastered by the change of feeling from the physical to the spiritual. Description—Emerging through the veil of Isis comes Death; overcome with horror I faint; awakening, fascination draws me closer; then with exhilarated joy I recognize Death as but the Gateway of Heaven.

Seven Exercises

Wave Lines 〰〰 Emotional.

Exercise No. 1. The Wave Line Balance.
Arm positions.
1st to 2 to R. 3d wave / L. 3d lower   thus
3d to 1st to 2d to R. 3d lower / L. 3d wave   thus

Exercise No. 2. The Harmonic Three with Wave Lines.
Arm Positions.   1st to 2d to 3d to 1st   Do continuous.

Exercise No. 3. The Wave Lines Diagonal.
Both arms start at 3d left lower   thus   to 2d to 3d right upper   thus   return to starting point   thus

EXPLANATION.

1. THE WAVE LINE BALANCE.

Start with the standing poise. Pass arms through 1st to 2d position and then execute a Wave Motion with right arm as per trace line to 3d position, while left passes to 3d lower position, and is a counter balance. This is a Straight (right side) Attitude. The arms then move to 3d or balance position in order to go to 1st position. Now repeat same to left. Hands in beauty position and open at 3d position. The wrist action is strongly called for in this exercise. Repeat 4 times. Do alternate. Slow gavotte or waltz music.

2. THE HARMONIC THREE WITH WAVE LINES.

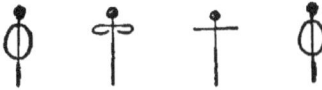

Standing poise. Move arms through the three positions as per cut, making a slow continuous movement, and add the Wave Motion, especially at the change from one position to the other. The head action and a slight Wave Motion of the body emphasizes the emotion displayed in this movement. Do continuous. 32 bars slow gavotte music.

3. THE WAVE LINES DIAGONAL.

Start and assume Attitude as per illustration. Both hands at left side. Pass arms with a wave motion (per trace line) to 3d upper on right side. As the arms cross the body diagonally upward, weight of the body is changed to lean to the right. (Technical—start with Straight (left side) Attitude and move to Straight (right side) Attitude. Return arms and body to starting point. Repeat 4 times. Now repeat exercise from right side—with arms both at right and body Straight (right side) Attitude. Repeat 4 times. Count 8 for each wave up or down. Slow gavotte music.

Exercise No. 4. The Upward Advance (A Pastoral.)

Subjects—Gathering Inspiration, Receiving Inspiration, Inspired.

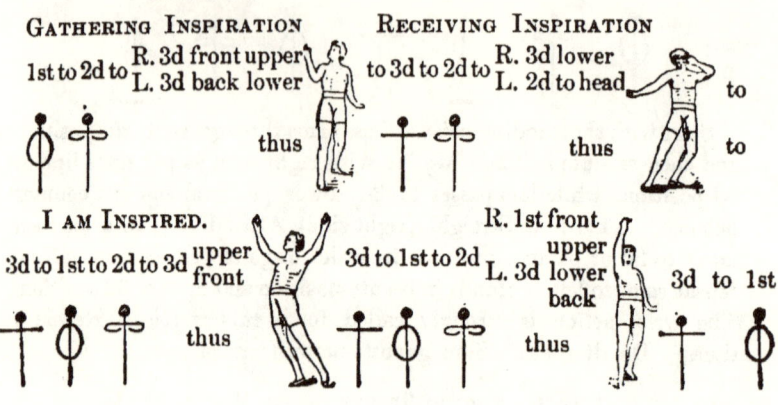

Gathering Inspiration
1st to 2d to R. 3d front upper
L. 3d back lower
thus

Receiving Inspiration
to 3d to 2d to R. 3d lower
L. 2d to head
thus to

I am Inspired.
3d to 1st to 2d to 3d upper front
thus

3d to 1st to 2d R. 1st front upper
L. 3d lower back
thus

3d to 1st

Exercise No. 5. The Emotion of Love.

Subjects—The Stars Above. To thee my Heart. At thy Feet.

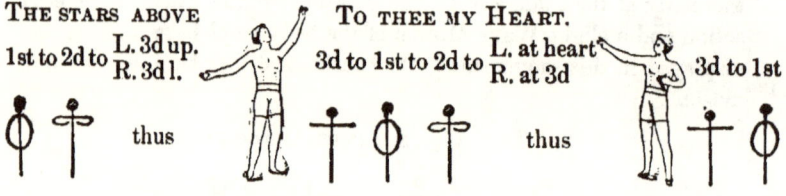

The stars above
1st to 2d to L. 3d up.
R. 3d l.
thus

To thee my Heart.
3d to 1st to 2d to L. at heart
R. at 3d
thus

3d to 1st

Repeat movements "*The stars above*" and "*To thee my heart ;*" finish in last attitude only kneeling thus

" At thy feet."

4. THE UPWARD ADVANCE (A PASTORAL.)

Assume with 1st arm movement, Straight (forward knee bend) Attitude, see 1st cut. The Wave Motion is diagonally upward in front. Start from a sinking movement of the body and go upward, diagonally forward. Accent this Attitude. Now (body natural,) move the arms through position to 2d Attitude (illustrated and accented.) This is an Arch (inward knee bend.) Attitude. Move the arms through next position, at same time (body natural—change weight to front foot) putting body in a preparatory position for 3d illustration (accented.) This is an Arch (outward) Attitude. From here (body natural) move arms and assume last Attitude. For the mental expression see No. 4 page 74. These Attitudes are strengthened by starting from a sinking position and then go up; also during the arm movements preparatory blendings of the poises should be observed so as not to make the assuming of any Attitude abrupt. Grace of motion is a continuous motion. Count to accent 4, hold accent 4. Same count until finished. Do 4 times. Slow gavotte music.

5. THE EMOTION OF LOVE.

Subjects—THE STARS ABOVE. TO THEE MY HEART. AT THY FEET.

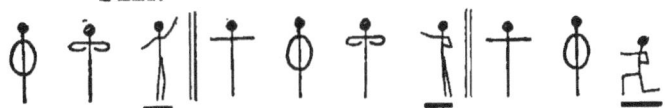

Assume Attitude with 1st arm movement. This is a Straight (left) Attitude. Accent this Attitude. With next arm movements (face right,) assume 2d Attitude. This is a Straight (right) Attitude, with left hand at heart, (accented.) Now repeat 1st arm movements and first Attitude (accented,) then 2d arm movements and 2d Attitude (accented.) In the position of this 2d Attitude slowly kneel down, see illustration, (accented.) Count to accent 4, hold accent 4; same count until kneeling, count 4. Do 4 times. Slow gavotte music.

EXERCISE NO. 6.   IN A GARDEN FAIR.

Subjects—HE CALLS TO HER, SHE IS IN A BALCONY. HIS PRESENCE THERE. HIS HEART TO HER. HIS LIPS TO HER. HIS WORSHIP. HIS PERPLEXITY.

HE CALLS TO HER, SHE IS IN A BALCONY.   HIS PRESENCE THERE.

both
3d left lower
thus

to 2d to 3d right upper
thus

return to
starting point
thus

HIS HEART TO HER.   HIS LIPS TO HER.

1st to 2d to L. at heart
R. at 3d upper
to 3d to 1st L. at heart.
R. to lips to 3d upper
thus
thus

HIS WORSHIP.   HIS PERPLEXITY.

3d to 2d Hands clasped
thus
to 3d to 2d to R. at chin
L. at 3d lower

EXERCISE NO. 7.   THE GATEWAY OF HEAVEN—DEATH.

Subjects—DISCOVERY AND IGNORANCE. CURIOSITY AND DAWNING. HORROR. REALIZATION. THE TRUTH—HEAVEN.

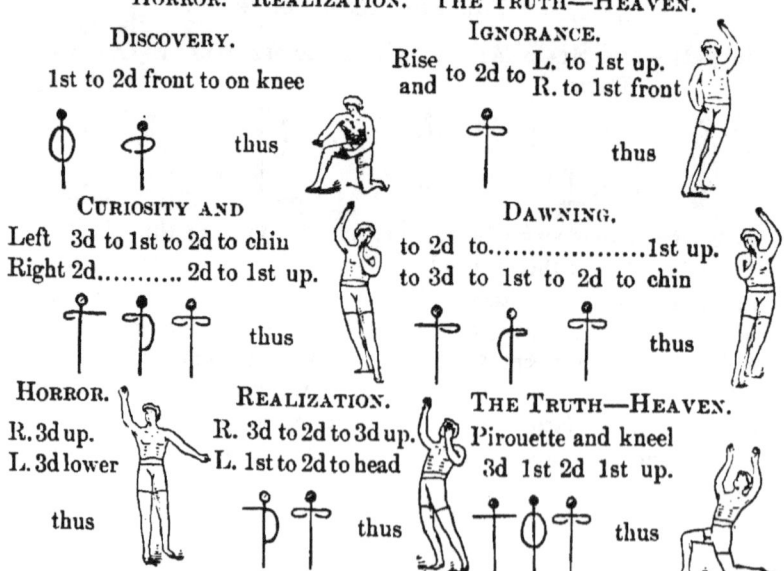

DISCOVERY.
1st to 2d front to on knee
thus

IGNORANCE.
Rise and to 2d to L. to 1st up.
R. to 1st front
thus

CURIOSITY AND
Left 3d to 1st to 2d to chin
Right 2d.......... 2d to 1st up.
thus

DAWNING.
to 2d to .... 1st up.
to 3d to 1st to 2d to chin
thus

HORROR.
R. 3d up.
L. 3d lower
thus

REALIZATION.
R. 3d to 2d to 3d up.
L. 1st to 2d to head
thus

THE TRUTH—HEAVEN.
Pirouette and kneel
3d 1st 2d 1st up.
thus

6. IN A GARDEN FAIR.

HE CALLS TO HER, SHE IS IN A BALCONY. HIS PRESENCE THERE. HIS HEART TO HER. HIS LIPS TO HER. HIS WORSHIP. HIS PERPLEXITY.

Start same Attitude and Arm Positions as No. 3, (Wave Lines Diagonal,) thus move both arms from left side diagonal up to right side; body moving from Left Poise to Right Poise (accented); this representing "*He calls to her.*" Now return arms to left side (body to Left Poise) and looking up, (accented) representing "*His presence there.*" Then (body natural) follow arm positions, start to blend the body into 4th Attitude Arch (right,) accented ; (¾ front view) with left hand on heart. Follow (body natural) arm positions and start to blend body to 5th Attitude Arch (right), accented ; the fingers of right hand nearly touch lips in passing in Wave Lines from 2d to 3d upper positions. Follow (body natural) arm positions and start to blend body to 6th Attitude Arch (right,) accented ; here the hands are clasped. Now move arms to last Attitude "*His perplexity*" the body blending to a Straight (left knee bend,) as per illustration (accented); the head here expressing the intention. Repeat from right side use reverse positions. Count to accent 4, hold accent 4 ; same count to finish. Use Trio of gavotte music.

7. THE GATEWAY OF HEAVEN—DEATH.

Subjects—DISCOVERY AND IGNORANCE. CURIOSITY AND DAWNING. HORROR. REALIZATION. THE TRUTH—HEAVEN.

Start Standing position, move arms and slowly kneel, (right foot in front,) and hold attitude per illustration, meaning "*Discovery.*" Rise slowly, move arms, and blend body to 2d Attitude (accented,) meaning "*unable to fathom;*" this is a Straight (back) knee bend. Follow (body natural) arm positions and start to blend body preparatory to assume 3d Attitude (accented) ; this is an Arch (left knee bend.) Now (body natural) move arms and blend body to assume 4th Attitude (accented); this is an Arch (right knee bend) ; the expression of the last two Attitudes mean "*Curiosity and Dawning.* Now step on right foot and assume a Straight (forward knee bend) ¾ front ; clinch hands ; the expression of head and shoulders is "*Horror.*" Remain with weight on right foot and assume arm position and an Arch (outward) Attitude (¾ front view) ; as per cut (accented,) meaning "*Realization.*" The last is the Pirouette, turning as if fainting and *fall.* (See Study of Flexibility) ; remain on floor 8 counts at *fall.* Rise on knees (left knee up,) and repeat entire ; reverse all movements and finish with Pirouette, and kneel as per last cut, meaning "*The Truth—Heaven;*" the mental is now purely spiritual and the face should express seeing the Gates of Heaven. Count 4 to accent, hold accent 4 ; same count until finished. Use slow gavotte music.

SPIRAL MOVEMENTS POWER.

1. THE SPIRAL MOTION.

The Spiral Motion is always upward. It embraces all the elements of the Circle lines, Wave lines and Curved lines. When the Spiral Motion is added to the movement of the body the sinking or bending of the knee is included. As a knee bend gives strength to a poise, so does the Spiral Motion give force to the curved lines. This motion calls for that intense expression of the Soul, and developes Personal Magnetism. This exercise is for the Spiral Motion of each arm, with the same motion added to the body.

2. THE CALL TO GOD.

Subjects—HIM. THE OFFER. THE CALL TO GOD. THE DOUBT. ANSWERED.

This exercise is entirely of a spiritual nature, except in one instance—"*The Doubt*" and that is but a thought of humility in a spiritual sense. Now the spiritual is the master of the physical. Opening with the first subject, "*Him*," the Spiral Attitude tells the whole story, if the interpreter has developed his Soul expression and is permeated with it. The eyes and the contour of poise should transform his physical for the time being. The attitude "*The Offer*," well expresses offering yourself to God. If the student has followed the foregoing outline, he will at the next subject "*The Call to God*" reach the climax of the exercise; the Spiral Motion and the Arch Poise are the strongest, and the inner expression is strengthened to its fullest extent by the store of Reserve Power gathered by the student. The attitude "*The Doubt*," has been explained. The last is the final climax "*Answered*," which is a pure Spiral Motion. See Study of Control of Will and The Forces, they will assist the purging of all carnality, which would be foreign here, and make the proper execution of this exercise impossible.

3. TAKE MY SOUL IN THY KEEPING.

Subjects—THE SOUL'S AWAKENING. SUPPRESSED SUSPENSE. MY SOUL IN THY KEEPING.

This exercise is of a similar nature to No. 2 except that the interpreter has developed to a certain extent, the forces within him, especially that of Personal Magnetism. With its proper conception, adding of course the exercises previously mastered, he will portray it properly. This exercise is more the expression of the Forces within than a mental photograph, although everything is registered at the brain. In the first subject, "*The Soul's Awakening,*" if the mind has been so charged, the feeling within would be as of the opening of a rose. Ideality would possess the mind and the Force would ascend to a point that to the keen student, all would be glory and sunshine, and as the mood changes to the next, "*Suppressed Suspense,*" the face is covered, tears are shed and the emotion sways the body by the masterful spiritual feeling, which has overcome the physical almost to the point of nonenity. These exercises are doubly described, that is motions of the body are given and also the thoughts that should engross the mind. As the thoughts here given surge through the mind the exterior cannot help express what is passing within. As the successful actor becomes for the time being incarnated in his part, so I insist that momentarily you must lose your identity and become but the living expression of the thoughts within.

4. THE WARRIOR.

Subjects—DETERMINATION. DEFIANCE. THE SWORD. THE CAUSE FOREVER.

The study here is heroic, and the movement therefore more accelerated. Determination and strong force of character are displayed and devotion to the physical sense of honor. They reach a spiritual intention at the attitude of "*The Cause Forever.*" The subjects are almost self explanatory, and when the student can lose "self" to that extent that his auditors almost feel his masterful presence, as a warrior "*Determined,*" he is bringing out that which is within him. The lesson is given here to study. The very attitude of "*Defiance*" should inspire the thought. The attitude of "*The Sword*" is only preparatory to the climax "*The Cause Forever,*" where the Spiral Motion of the swinging of the sword and the body motion should be most pronounced, which would be fitting for a leader of men or soldiers and carry all by the unseen Force he displays.

5. The Plea.

Subjects—Sorrow. A Plea. Divine Supplication. Prayer and Resignation.

This is a study in opposition to that of No. 4. The titles here clearly define the impersonations. They are a combination of the sad emotions changing to that of a spiritual nature. The sorrowful attitude first assumed, will well express the miserable state or mental condition desired and can be intensified to the extent of the ability of the student. A plea for deliverance follows, which leads to the climax "*Divine Supplication.*" The slow retiring movement to "*Prayer*" is a most excellent study. The last, "*Resignation,*" although the action may seem unimportant, is one of the "leading" studies in this work, that of Flexibility and Relaxation.

6. An Allegory.

Subjects—Attention. Telling a Multitude. There appeared and was seen a vision. Surprise. Of one on high. I only a Mortal.

This is a study of Reserve Power. The student describing his emotions of having seen a vision. The first movement of slowly raising the hand, should be backed by all the Force in reserve meaning to get the "*Attention*" and silence of the multitude. The next study "*Telling a Multitude*" is a sweeping arm motion, impressed so strongly with the head as to penetrate the most remote corner. The narration is now unfolded to the audience and is graphically described, "*There appeared and was seen a vision,*" with which the final expression changes. The mood rapidly changes again to "*Surprise.*" Continuing to describe his emotions, the Spiral Motion reaches its highest upward stretch in expressing "*Of one on high.*" The last "*I only a Mortal*" is a study of relaxation from finger tips to the toes.

7. THE ARABESQUES.

The Arabesques relate to the Mercury Balance Position and are taken from the statue of Mercury, and by ancient artists were favorites for panel drawings. Blasis in speaking of Arabesques called them improper attitudes (but favored them) because he could not locate their gravity. The Laws of this work, cover the points in bending, in the case of counter balance at the control of gravity centers. They certainly are very beautiful if the accent is given at the pose of Mercury on the Right or Left Arch Poise as marked.

Seven Exercises. Spiral Movement  Power.

Exercise No. 1.   The Spiral Motion.

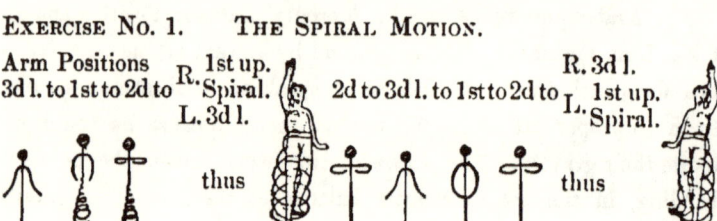

Arm Positions
3d l. to 1st to 2d to R. 1st up. Spiral. L. 3d l. thus   2d to 3d l. to 1st to 2d to   R. 3d l. L. 1st up. Spiral. thus

Exercise No. 2.   The Call to God.

Subjects—Him. The Offer. The Call to God. The Doubt.
        Answered.

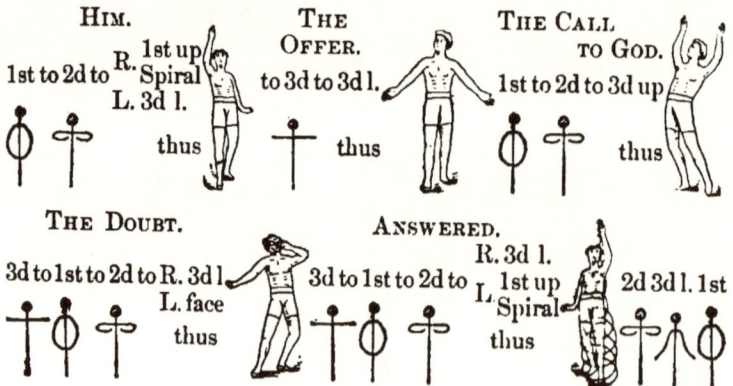

Him.
1st to 2d to R. 1st up Spiral L. 3d l. thus

The Offer.
to 3d to 3d l. thus

The Call to God.
1st to 2d to 3d up thus

The Doubt.
3d to 1st to 2d to R. 3d l. L. face thus

Answered.
3d to 1st to 2d to R. 3d l. L. 1st up Spiral thus   2d 3d l. 1st

EXPLANATION.

1. THE SPIRAL MOTION.

Begin this exercise from a Sinking (knee bend) Attitude. Move the arms through the positions, at same time executing a Spiral Motion of the body upward, and after the arms reach 2d position (the first time) the right arm continues the Spiral upward to its highest stretch; the left hand going to 3d lower is a counter balance, (see 1st illustration.) From this Attitude to begin the exercise over again, you must dissolve gracefully by having the arms and body return to a natural Attitude; this time the left arm finishes the Spiral upward, (see 2d illustration.) The eyes follow the hand up. The Spiral of the arms must be observed. The body assumes a slight Arch (outward) at Attitudes that are illustrated and accented. Count 8 for Spiral Motion upward and same to recover. Repeat 4 times. Slow gavotte music.

2. THE CALL TO GOD.

Subjects—HIM. OFFER. CALL TO GOD. DOUBT. ANSWERED.

Begin the first subject in same manner as No. 1, to illustration, accented. Then move arms through next position, and assume a Straight (back knee bend) Attitude, and look up as if "*Offering yourself to God.*" Follow the next arm movements (body natural) and by a Spiral Body Motion assume the 3d Attitude, (accented); this is an Arch (outward) Attitude, (¾ front view,) and is easily assumed by placing weight on front foot. Move arms through next positions (body natural) then move body preparatory to assume 4th Attitude, (illustrated and accented); this is a Straight (left side knee bend) Attitude, the left hand crosses face from 2d position. Now follow arm and body movements the same as No. 1, (with this difference, use left hand for Spiral)—meaning "*Answered.*" Recover a natural poise by following the last positions (per cut) of the arms. Count to accent 4, hold accent 4; same count until finished. Repeat 4 times. Use an appropriate hymn and sway to its rhythm.

EXERCISE No. 3.   TAKE MY SOUL IN THY KEEPING.

Subjects—THE SOUL'S AWAKENING.   SUPPRESSED SUSPENSE.   MY SOUL IN THY KEEPING.

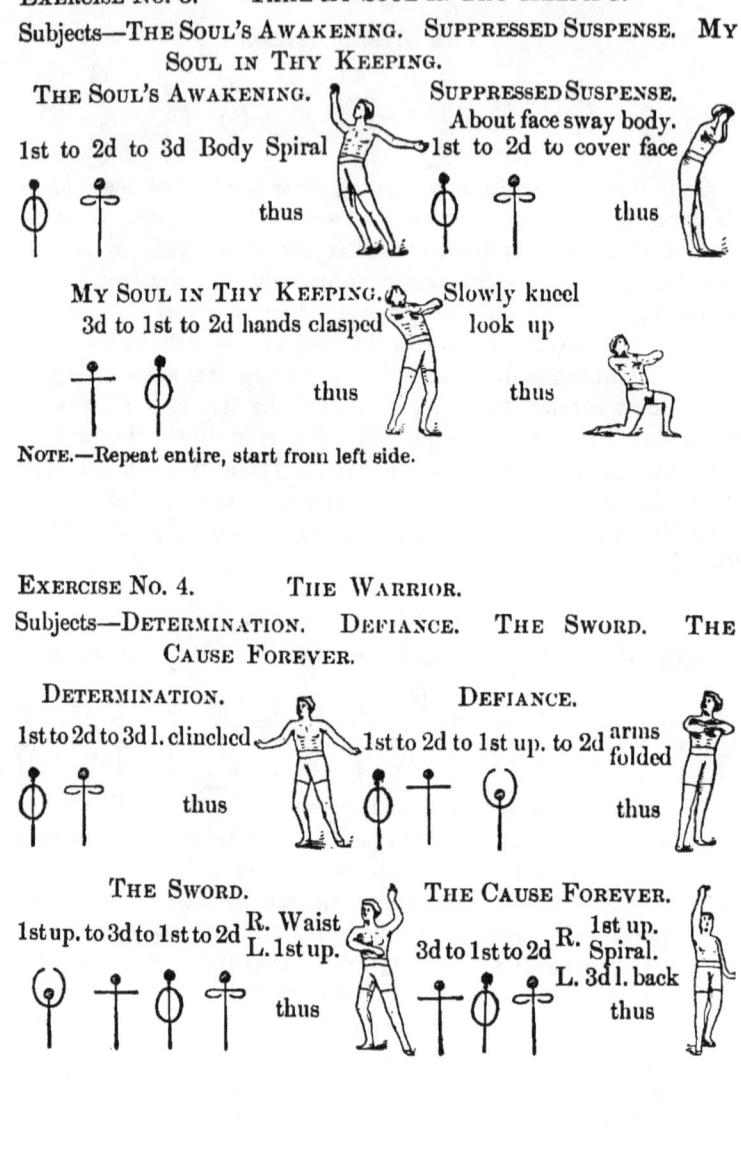

THE SOUL'S AWAKENING.
1st to 2d to 3d Body Spiral
thus

SUPPRESSED SUSPENSE.
About face sway body.
1st to 2d to cover face
thus

MY SOUL IN THY KEEPING.
3d to 1st to 2d hands clasped
thus

Slowly kneel
look up
thus

NOTE.—Repeat entire, start from left side.

EXERCISE No. 4.   THE WARRIOR.

Subjects—DETERMINATION.   DEFIANCE.   THE SWORD.   THE CAUSE FOREVER.

DETERMINATION.
1st to 2d to 3d l. clinched
thus

DEFIANCE.
1st to 2d to 1st up. to 2d arms folded
thus

THE SWORD.
1st up. to 3d to 1st to 2d R. Waist L. 1st up.
thus

THE CAUSE FOREVER.
3d to 1st to 2d R. 1st up. Spiral. L. 3d l. back
thus

3. Take my Soul in Thy Keeping.

Subjects—The Soul's Awakening. Suppressed Suspense. My Soul in Thy Keeping.

Begin the Spiral Motion of the body with arm movements to 1st Attitude (accented); this is an Arch (outward) Attitude (¾ front view) facing right. Recover (body natural) and at same time about face (to left); pass arms through next positions and assume the 2d Attitude, by covering the face, swaying the body back and forth. Now recover (body natural) and start the arm movements with a Spiral Motion of the body to the 4th Attitude; this is an Arch (outward) Attitude, (¾ front view,) facing left. Now slowly slide right foot backward and kneel (eyes looking up.) Repeat by rising and blending in the first Attitude, face left, etc. Do 4 times. Use an appropriate hymn and sway to its rhythm.

4. The Warrior.

Subjects—Determination. Defiance. The Sword. The Cause Forever.

Start from standing poise, step forward on right (weight on right foot,) moving arms through positions to 1st Attitude; this is a Straight (forward knee bend) Attitude, accented and ¾ front view, with hands clinched and "*Determination*" expressed in every feature. Recover (body natural) and blend arms through next positions, and when at 1st upper let them down and fold them; by placing weight on left foot assume 2d Attitude; this is a Straight (back) Attitude, and is accented. Next undo the arms per arm positions (body natural) and continue arm movements and blending of body, by placing weight on right foot assume 3d Attitude; this is a Straight (forward knee bend) Attitude; ¾ front view; after leaving 2d position have the right hand resting on the hilt of an imaginary sword and the left at 1st upper, completing the picture. Finally you recover and do a purely Spiral Motion of the body and arms, the right hand leads and denotes the waving of the drawn sword. Count to accent 4, hold accent 4; same count until finished. Repeat entire 4 times. Slow gavotte music.

EXERCISE No. 5.   THE PLEA.

Subjects—SORROW. A PLEA. DIVINE SUPPLICATION. PRAYER AND RESIGNATION.

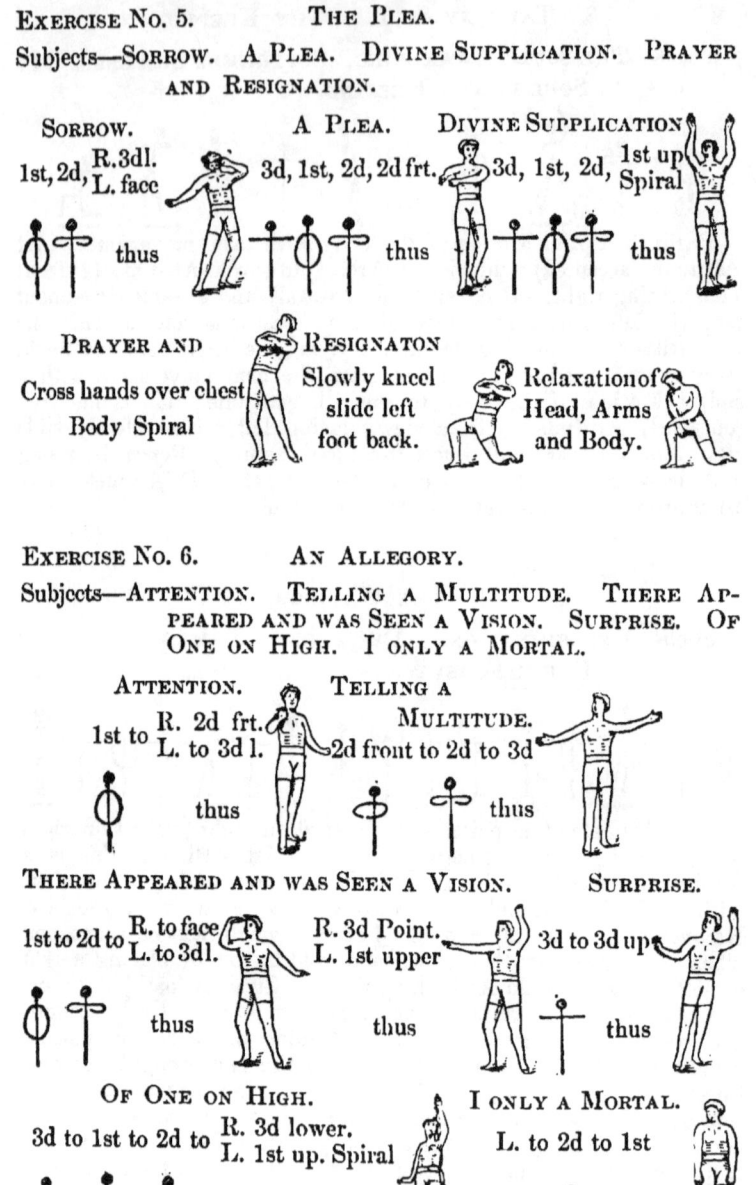

SORROW.
1st, 2d, R. 3dl.
L. face
thus

A PLEA.
3d, 1st, 2d, 2d frt.
thus

DIVINE SUPPLICATION.
3d, 1st, 2d, 1st up Spiral
thus

PRAYER AND
Cross hands over chest.
Body Spiral

RESIGNATON
Slowly kneel slide left foot back.

Relaxation of Head, Arms and Body.

EXERCISE No. 6.   AN ALLEGORY.

Subjects—ATTENTION. TELLING A MULTITUDE. THERE APPEARED AND WAS SEEN A VISION. SURPRISE. OF ONE ON HIGH. I ONLY A MORTAL.

ATTENTION.
1st to R. 2d frt.
L. to 3d l.
thus

TELLING A MULTITUDE.
2d front to 2d to 3d
thus

THERE APPEARED AND WAS SEEN A VISION.
1st to 2d to R. to face
L. to 3d l.
thus

R. 3d Point.
L. 1st upper
thus

SURPRISE.
3d to 3d up
thus

OF ONE ON HIGH.
3d to 1st to 2d to R. 3d lower.
L. 1st up. Spiral
thus

I ONLY A MORTAL.
L. to 2d to 1st
thus

5. THE PLEA.

Subjects—SORROW. A PLEA. DIVINE SUPPLICATION. PRAYER AND RESIGNATION.

Start by placing weight on left foot behind, pass arms through arm positions and assume 1st Attitude; this is an Arch (inward knee bend) Attitude and is accented. Next recover position, passing arms through the position per cut, and by the head action assume 2d Attitude (weight on right foot) and accent it strongly by an inward emotion of pleading. Now the arms move to 3d and 1st positions, the body at same time sinking in order to execute a Spiral Motion to assume 3d Attitude, (weight on right foot) eyes looking up; this is an Arch (outward) Attitude, and is accented. Next hold this Attitude, let hands down and cross them over breast (eyes looking up,) slowly kneel by sliding left foot back; when kneeling lean back. The last is a collapse by relaxing every muscle and drooping forward. Count 4 to accent, hold accent 4; same count until finished. Repeat 4 times. Slow gavotte music.

6. AN ALLEGORY.

Subjects—ATTENTION. TELLING A MULTITUDE. THERE APPEARED AND WAS SEEN A VISION. SURPRISE. OF ONE ON HIGH. I ONLY A MORTAL.

Begin with 1st Attitude which is a Straight (forward) Attitude weight on right foot. The action of impressively raising the right arm and straightening of fingers should, with a proper flow of Magnetism (through the fingers) express this 1st Attitude. Now follow arm movements and transfer weight to left foot, (lean back) and express by the action of head and sweep of the arms the 2d Attitude; this is a Straight (back) Attitude, accented. Recover (body natural) and pass arms through positions, blend the body to the 3d Attitude, which is a Straight (right side knee bend) Attitude, accented, with the hand shading the eye. Now keep this same Attitude and point right hand, the left going to 1st upper, making the 4th Attitude. Next blend arms and body to 5th Attitude "*Surprise*" by placing weight on left foot behind; this is a Straight (back knee bend) Attitude. Recover and with arm position and body do a Spiral Motion (left hand leading) assume 6th Attitude. The last is an entire collapse or relaxation per illustration. Count 4 to accent, hold accent 4; same count until finished. Repeat, starting with left hand. Slow gavotte music.

EXERCISE No. 7.

THE ARABESQUES.—From Mercury Balance Poise.

Body in Arch (left) Attitude, finishing in Arch (right) Attitude.
Arm positions.

2d to 1st to 3d to 1st upper to 3d to 1st to 2d to R. 1st upper
L. 3d lower

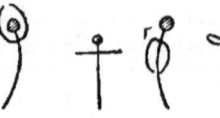 thus

Body in Arch (right) Attitude, finishing in Arch (left) Attitude.
Arm positions.

1st up. to 2d to 1st to 3d to 1st up. to 3d to 1st to 2d, R. 3d l.
L. 1st up

 thus

7. THE ARABESQUES.

Assume 1st arm positions, with an Arch (left side) Attitude, same time doing arm movements per cut; at 3d position body blending through natural body poise to an Arch (right side) Attitude, and assume Mercury poise, per illustration. Continue with body in this Arch Attitude (the face to front) and pass arms through positions per cut until 3d position, here blend body through natural poise to an Arch (left side) Attitude, and assume Mercury Poise per 2d illustration. These movements are a continuous blending and held only at accents. To continue from last Attitude the lower hand is passed to 1st upper as per cut; from this position return to beginning, repeat entire, making one continuous blending. Observe Law of Head and body facing front. This exercise embraces the Spiral Motion with Arch poising and presents this outline from any view. Do 4 times through. Use 64 bars slow gavotte music.

Seven Laws For The Curved Lines.

1. The Palm of Hands.

In Curved Lines the palm always turns at 3rd position,—down, when going downward, and up, when going upward.

2. The Eyes.

The eyes always follow the hand or hands, when they go above the line of vision. The eye always follows the highest hand.

3. The Head.

The head is always away from the raised hand (in the study of Grace,) and the head is in harmony with the active foot, (not the standing foot.) See Laws of Opposition.

4. The Balance Position.

The hands, just before the finish of the movement must come to a Balance Position, before going to a finish or end position. To change the movement to opposite side, be sure the hands come to a Balance Position before starting again. Any position, where the hands are in the same position, is a Balance Position. The blending of one movement into another is thus established.

5. Grace of Movement.

Positions that start at 1st position are given a better touch of Grace to start at 3d lower and then to 1st position.

6. Counter Poise of Arms.

Counter poise of the arms or hand, is a balance in a diagonal line; the hands doing same should never droop.

7. The Mental Expression of the Arm.

The thought expressed by the mind is usually directed to the mental of the arm which is the hand; therefore, the fingers and hand should give the intelligence of the movement.

Flexibility.

FLEXIBILITY.

1. Flexibility is an emotional adjunct. Emotion does not exist without flexibility.

2. FLEXING.

It is well to do the Flexing movements in a rotary manner, to inculcate a curve, but it is not necessary.

3. THE LAW.

The law for Flexing is to withdraw the *will power* or mental thought of strength from the part being flexed.

4. CONTROL OF SEPARATE PARTS.

The use of Flexibility or Relaxation of the different parts separately is to have such control of these parts, that no one part will ever do the natural work of any other.

5. VITAL FORCE.

The saving of vital force, or ease of motion is the perfection of using only such muscles as are required to execute such a motion, leaving those not necessary in repose. This is understood as "no waste" of vital force.

6. Relaxation.

Relaxation is rest, or release of tension. The relaxation of sleep restores those nervous forces, which are overtaxed during the waking hours. Physical exercise should not be continued, so as to overtax the strength.

7. Relaxation in Falling.

Flexing in Falling is one of nature's best tonics or invigorators for sluggish minds, as the reaction after Flexing is one of buoyancy of body and activity of mind. In the Falling Exercise the Relaxation starts at the base and the muscles relax upward as the body sinks to the ground. The position after falling should be—the top foot extended and the under foot crossed; *so when rising* the under hand is only used to assist in rising to a knee position (of the under foot.) From there, to an upright position, is simply rising.

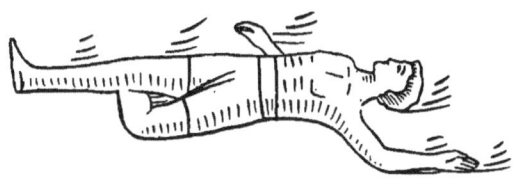

FLEXING EXERCISES.

1st EXERCISE
FLEX HANDS FROM WRIST.

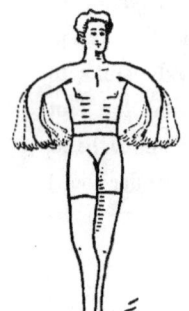

2d EXERCISE
FLEX FOREARMS
FROM ELBOW

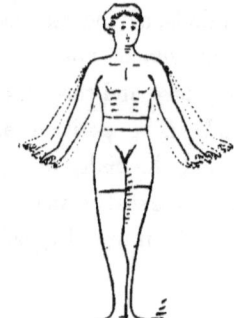

3d EXERCISE
FLEX ARM
FROM SHOULDER.

4th EXERCISE
FLEX FOOT
FROM ANKLE

5th EXERCISE
FLEX SHANK AND F(
FROM KNEE

6th EXERCISE
FLEX LEG
FROM HIP.

7th EXERCISE
FLEX ENTIRE
BODY.

EXPLANATION.

1. FLEX HANDS FROM WRIST.

Assume position of hands as per cut, drooping them loosely from wrist, and shake (or what is here called flexing) them. The *will power* is withdrawn from the parts being flexed. Use 16 bars of tremulo music.

2. FLEX FORE-ARMS FROM ELBOW.

Assume position of fore-arms as per cut, drooping them loosely from elbow and flex them. Same action as No. 1. Same music.

3. FLEX ARM FROM SHOULDER.

Assume position of right arm as per cut, drooping loosely from shoulder and flex it. Same action as No. 1. Repeat with left arm. Same music as No. 1 for each arm.

4. FLEX FOOT FROM ANKLE.

Assume position of right foot as per cut, drooping loosely from ankle and flex it. Same action as No. 1. Repeat with left foot. Same music as No. 1 for each foot.

5. FLEX SHANK AND FOOT FROM KNEE.

Assume position of shank and foot as per cut, drooping loosely from knee and flex it. Same action as No. 1. Repeat with left foot. Same music as No. 1 for each.

6. FLEX LEG FROM HIP.

Assume position of right leg as per cut, drooping loosely from hip and flex it. Same action as No. 1. Repeat with left leg. Same music as No. 1 for each leg.

7. FLEX ENTIRE BODY.

Simply withdraw will power from the entire body and shake yourself until you are limber all over, so that the resistance of any muscle has been eliminated. Same music, 16 bars.

Relaxation Exercises.

1st Exercise
THE ARMS
(PERPENDICULAR.)

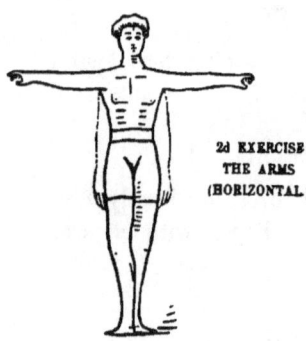

2d Exercise
THE ARMS
(HORIZONTAL.)

3d Exercise
THE LEG.

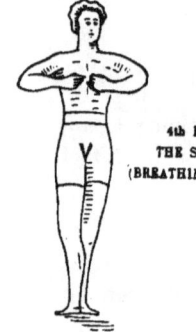

4th Exercise
THE SHOULDERS
(BREATHING—COMFORT.)

5th Exercise
THE NECK
(FLEXIBILITY.)

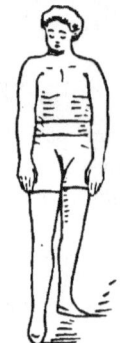

6th Exercise
RESTING
SLEEP

EXPLANATION.

1. THE ARMS. (PERPENDICULAR.)

Assume position as per cut by slowly raising right arm to 1st upper and suddenly withdraw will power of arm, allowing it to drop as if disjointed. To do this relax the shoulder to elbow, then wrist, in quick succession. The arm should pass close to the body. Repeat with left arm. Repeat with both arms. Do each two times. Count to assume position 6, and drop 7, 8. Slow schottische time.

2. THE ARMS. (HORIZONTAL.)

Assume position as per cut by slowly raising arms to 3d position, and suddenly withdraw will power of the arms, leaving them drop. Relax all parts of the arms at same time. Same music and count as No. 1. Repeat 4 times.

3. THE LEG.

Assume position as per cut by slowly raising right leg and suddenly withdraw will power, let it drop. Repeat with left leg. Do four times each leg. Count to raise 3, and drop on 4th count; now rest 4 counts. Slow schottische music.

4. THE SHOULDERS. (BREATHING-COMFORT.)

Place hands lightly on chest and inhale (expand chest.) Now slowly exhale, drop hands, and release the tension of shoulders and "set" them by relaxation until there is a feeling of comfort in the shoulders. This shoulder adjustment is a very important movement for the gravity of the body in any poise. This exercise is for relaxation of shoulders and chest expansion. Repeat 4 times. Gavotte music. For Breathing Exercise, see ¶5, page 31.

5. NECK. (FLEXIBILITY.)

Relaxation of neck is the same as Neck Exercise in physical corrections. Drop head in front and let it dreamily circle on an imaginary axis. Repeat three times each way. Slow waltz music.

6. RESTING SLEEP.

Standing position. Close the eyes and breathe regularly, releasing the tension of every muscle. Thus the will controls relaxation and in time would force sleep. This exercise requires the relaxation of the Mental Forces as well as the Physical. It will act as a tonic at any time for exhausted nerve force, replenishing the waste at once. Use low music. Do the same at 2 minute intervals.

7. FALLING AND RECLINING.

Start at standing poise, with relaxations of muscles under control of will. Slowly kneel down and extend right foot as per cut, and gradually lie down as per cut, (dotted line.) The under foot is crossed when down. To arise turn over on left side, place palm of left hand on floor, rise to a kneeling position, then to an upright one. Now repeat, kneel and extend left foot. This exercise is preparatory to Falling Exercise which follows on next page. Repeat four times. Use tremulo music.

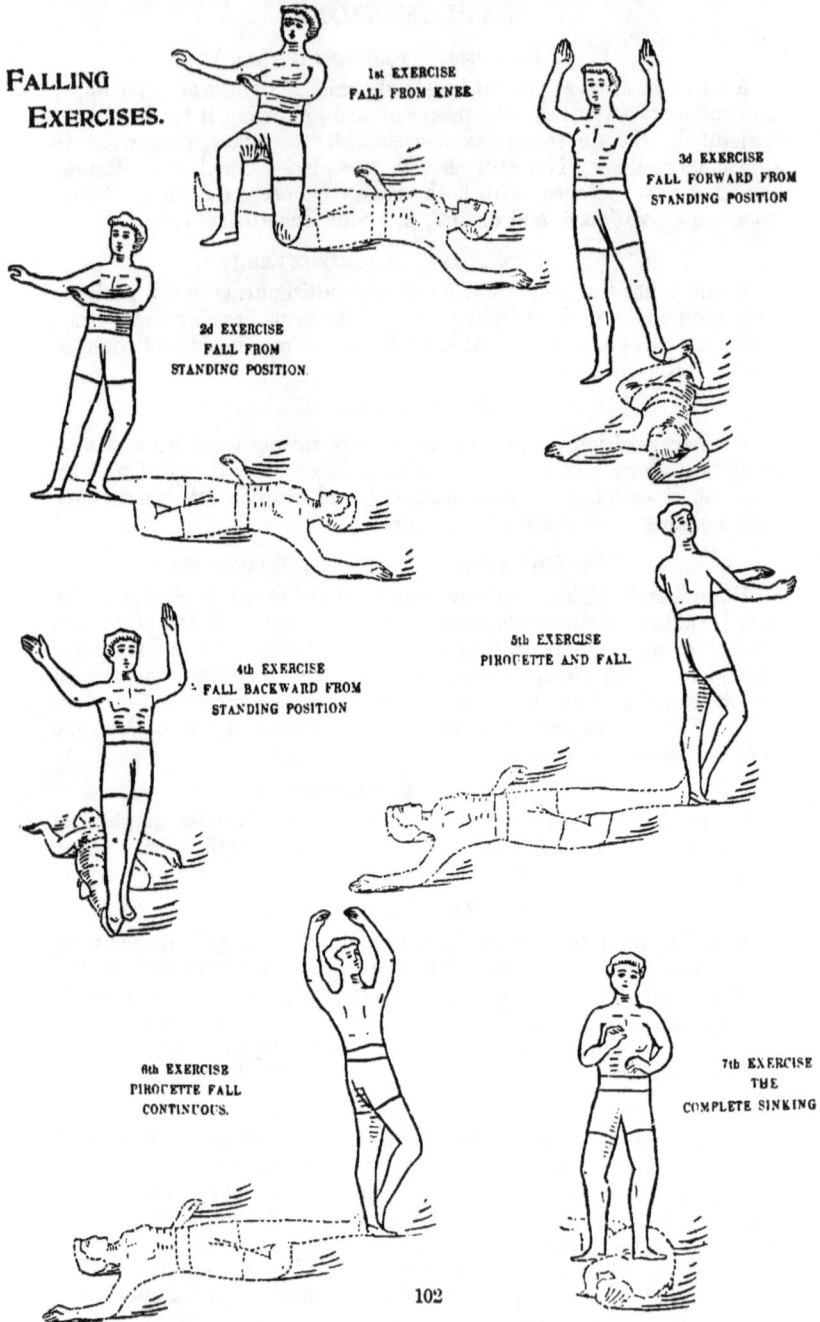

EXPLANATION.

1. FALL FROM KNEE.

Kneel and assume position as per cut, and relax in quick succession from knee to hip, to shoulder, to head and entire arm, and *fall* per dotted outline in cut. The resistance power should be withdrawn from each part as the body drops, and is one of the easiest exercises, if the will controls the relaxation; (see Flexing Exercises.) *Always fall suddenly as if shot.* Rise and repeat to left. Do 4 times. Hold or rise to position, count 6, and *fall* on count 7, 8. When proficient count 3 and *fall* on 4th count. Gavotte music.

2. FALL FROM STANDING POSITION.

Assume attitude as per cut, with weight of body on right foot. Now relax in quick succession (as the body drops) from ankle to knee, to hip to shoulder, to head to arms, and *fall* to right per dotted outline. Same principle as No. 1. Rise, per No. 7, (page 101,) and repeat, *fall* to left. Same count and music as No. 1. NOTE:—If timidity is shown two persons can join hands and fall alternately, the one assisting the other when *falling*.

3. FALL FORWARD FROM STANDING POSITION.

Assume attitude as per cut with weight of body on front (right) foot. Now relax in quick succession (as the body drops) from ankle to knee, to hip to chest, to head to arms, and *fall forward*, per dotted outline. To rise, turn on left side, and arise per No. 7, (page 101.) Same count as No. 1. Same principle and same music as No. 1.

4. FALL BACKWARD FROM STANDING POSITION.

Assume attitude as per cut with weight of body on back (left) foot. Now relax same as No. 3 and *fall backward* per dotted outline. Recover same as No. 3. Same principle, same count, same music as No. 1.

5. PIROUETTE AND FALL.

This is to Pirouette to right and then *fall* to the right. To Pirouette is to step to right, cross left toe over right foot, turn yourself around; then *fall* same as No. 2, or per dotted outline. Rise per No. 7, (page 101,) and repeat entire to left. Count 2 bars to Pirouette; 2 bars to *fall*, and 4 to recover to standing position. Waltz music. If possible, use arm movement in Pirouette. Do four times each.

6. PIROUETTE FALL, CONTINUOUS.

Pirouette to right and then immediately *fall* to right by letting right foot slide underneath you—the entire body relaxing instantly. Rise instantly and Pirouette to left and *fall*. Do alternate—continuous. Count 1, 2, 3 to Pirouette, and 1, 2, 3 to *fall* and rise. Polka time. This exercise, if executed correctly, shows perfection of Flexibility and Relaxation.

7. THE COMPLETE SINKING.

Standing position. Do Flexibility Exercise 7, (page 99.) Tremulo of whole body and collapse gradually down as per outline in cut. Tremulo music. Do four times.

ATTITUDES.

ATTITUDES.

An Attitude is the body posed or poised in any position. To pose the body in an Attitude is to assume any stationary posture. Poise of the body or parts of the body is to assume any balance Attitude wherein the laws of opposition are observed.

In the Straight Attitudes the mental predominates; in the Arch Attitudes the emotional predominates; in the Bending Attitudes the physical predominates. When the intent is combined with these Attitudes of the body, the mental, emotional or physical may relate to Gravity, Flexibility or Force; or to the Mind, Soul and Body.

The Study of Grace in Attitudes is the continuous blending of one Attitude into another. An Attitude reaches its proper Law of Opposition and Harmony at the accented period, or when it is stationary for the time being.

ATTITUDES.

Consist of 1. Straight Attitudes—Mental—Gravity.

2. Arch Attitudes—Emotional—Flexibility.

3. Bending Attitudes—Physical—Force.

DERIVATIONS.

4. Knee Bend Attitudes.

5. Beyond the Base Line.

6. Kneeling and Lying.

7. Poise.

STRAIGHT ATTITUDES.

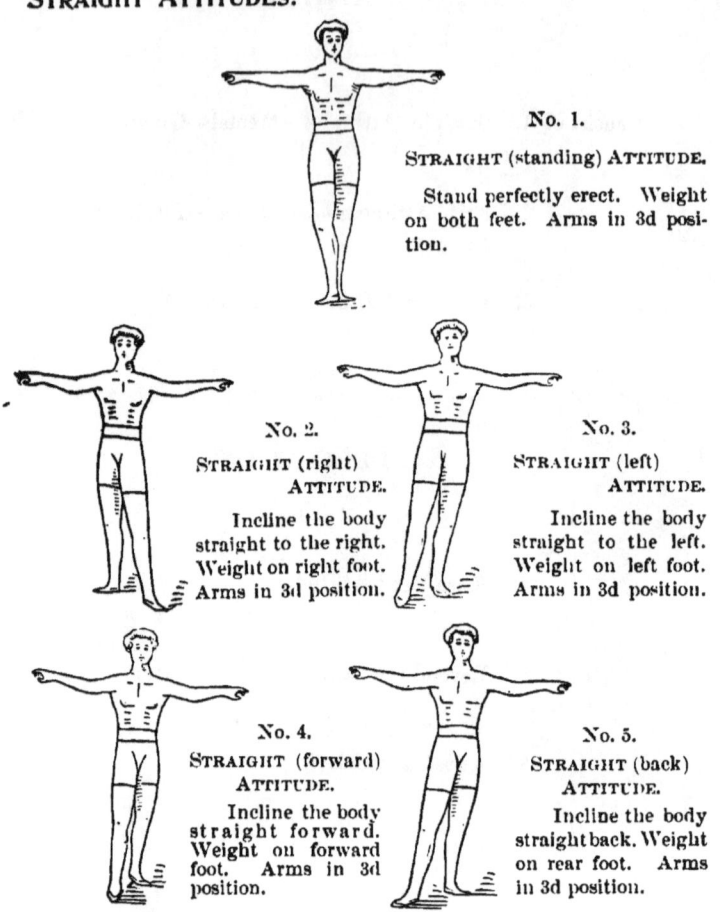

No. 1.
STRAIGHT (standing) ATTITUDE.
Stand perfectly erect. Weight on both feet. Arms in 3d position.

No. 2.
STRAIGHT (right) ATTITUDE.
Incline the body straight to the right. Weight on right foot. Arms in 3d position.

No. 3.
STRAIGHT (left) ATTITUDE.
Incline the body straight to the left. Weight on left foot. Arms in 3d position.

No. 4.
STRAIGHT (forward) ATTITUDE.
Incline the body straight forward. Weight on forward foot. Arms in 3d position.

No. 5.
STRAIGHT (back) ATTITUDE.
Incline the body straight back. Weight on rear foot. Arms in 3d position.

Execute Nos. 4 and 5, at a ¾ angle front view.

PRACTICE.

Arm Movements No. 1. The Harmonic Three (Circle Series, pg. 70) with Attitudes Nos. 2 and 3.

Arm Movements No. 2. The Complete Circle (Circle Series, pg. 70) with Attitudes Nos. 4 and 5.

Arm Movements No. 3. The Alternate Exercise (Circle Series, pg. 70) with Attitudes Nos. 2, 3, 5 and 4.

Do the above continuous, blend from one Attitude to the other; using Arm Movements at same time. At last Attitude (No. 4) bring arms to 3d position then to 1st position. Gavotte music.

EXPLANATION.

Assume each Attitude in the order given until familiar with them. Observe laws of carriage of the body and rules here given to assume Straight Attitudes.

The Straight Attitudes are so named because the body is in a straight line from base to head, either erect or inclined.

The 3d arm or balance position is a beneficial arm movement to use in conjunction with assuming these Straight Attitudes.

The weight of the body should always be on the outside foot; the other foot just far enough away to secure a perfect equilibrium. To illustrate: in the Straight (right) Attitude the weight is on the right foot; Straight (left) Attitude, weight is on left foot; Straight (forward) Attitude, weight is on forward foot; Straight (back) Attitude, weight is on rear foot. (See Heel, page 26.)

The Straight (standing), Straight (forward) and Straight (back) Attitudes, may be assumed at an angle $\frac{3}{4}$ view front, which increases their effectiveness and should be assumed whenever possible.

The arm movements retain their same position at the same angle or bend, that the body assumes in any Attitude, which establishes a Harmony and Opposition of the head and acting foot and waist. (See study of Harmony and Opposition.)

The foundation of Gravity of Poise is fully established in Straight Attitudes; *i. e.* the gravity line from neck to heel of standing foot.

ARCH ATTITUDES.

No. 1.

ARCH (right) ATTITUDE.

Assume an Arch of the body with the right side, (per cut.) Face and body front view. Weight on right foot. Arms in 1st upper position.

No. 2.

ARCH (left) ATTITUDE.

Assume an Arch of the body with the left side, (per cut.) Face and body front view. Weight on left foot. Arms in 1st upper position.

No. 3.

ARCH (outward) ATTITUDE.

Assume an Arch of the body outward (back concave.) Weight on forward foot. Face and body front view. Arms in 1st upper position.

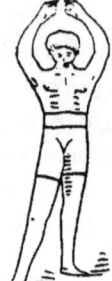

No. 4.

ARCH (inward) ATTITUDE.

Assume an Arch of the body inward (back convex.) Weight on rear foot. Face and body front view. Arms in 1st upper position.

Execute Nos. 3 and 4 at a ¾ angle front view.

PRACTICE.

Arm Movements No. 2.   The Complete Circle (Circle Series pg. 70) with Attitudes Nos. 1 and 2.

Arm Movements No. 3.   The Alternate Exercise (Circle Series pg. 70) with Attitudes Nos. 1 and 2.

Arm Movements No. 2.   The Complete Circle (Circle Series pg. 70) with Attitudes Nos. 3 and 4.

Do the above continuous, blend from one Attitude to the other; using Arm Movements at same time. Gavotte Music.

EXPLANATION.

Assume each Attitude in the order given until familiar with them. Observe laws of carriage of the body and rules here given to assume Arch Attitudes.

The Arch if continued would form part of a circle from base to top; that is from standing heel along the outside of Arch to the highest hand or hands.

The 1st position upper is a beneficial arm movement to use in conjunction with Arch posing.

In executing an Arch pose the weight of the body should be on the outside of the Arch; that is—Arch (right) Attitude, the weight is on right foot, etc., the same as in Straight Attitudes. The face and entire body presenting a front view.

In the study of Harmony and Opposition, the head is in harmony with the acting foot; *not the standing foot.* The waist is opposed at any point (around the waist) where this harmony is established. The head is then in opposition to the highest hand, *i. e.* inclined from the high hand.

The outward and inward Arch Attitudes may be assumed at an angle ¾ view front, giving them a dramatic effect, and should be assumed whenever possible.

Flexibility is necessary to the perfection of Arch posing. Harmony of poise of the outside muscles of the Arch will give correct curves. Practice before a glass will be of great assistance.

BENDINGS.

Consist of Knee Bends with Straight Attitudes.

Knee Bends with Arch Attitudes.

Spinal Column (bend) Attitudes.

STRAIGHT (knee bend) ATTITUDES apply to all Straight Attitudes (page 108.) They are similar, with this addition—the knee of the standing foot is bent. In Knee Bend Attitudes the weight of the body is nearly equal on both feet, but stronger on the leg of which the knee is bent.

ARCH (knee bend) ATTITUDES apply to all Arch Attitudes, giving the inward and outward Arch Attitudes a special significance. These Attitudes are the same as Arch Attitudes (page 110) having this addition—the knee of the standing foot is bent. The weight of the body is the same as in Straight Attitudes.

SPINAL COLUMN (bend) ATTITUDES are Straight Attitudes bending the body backward. The weight of the body is on rear foot, which forms the base.

EXEMPLIFICATION.

STRAIGHT (forward knee bend) ATTITUDE.

This is simply a Straight (forward) Attitude with knee of rear leg bent. The arm positions are—right arm 1st upper, left arm is a counter poise 3d lower behind. The counter part of this Attitude is a Straight (back knee bend) Attitude. The sense in either is that of an heroic order, such as "defiance," "the onslaught," "lead on," "away to war," "to God above," "justice," "righteousness," and similar subjects. The weight of the body is nearly divided on both feet with a stronger inclination on the bending leg.

ARCH (inward knee bend) ATTITUDE.
¾ front view.

This is an Arch (inward) Attitude, with the addition of the knee bent of the outside leg of the Arch, which is the rear foot. This Attitude is usually executed at an angle ¾ view front. The sense that is expressed in this Attitude is "sorrow," "fear," "horror," "cowardice," "curiosity," and kindred subjects. The facial expression with the arm and head movements denote the emotion intended.

SPINAL COLUMN (bend) ATTITUDE.
Beyond the Base Line.

This is a side view of the body. When this Attitude has been assumed, the view is of the right side and face is to right. The feet are separated, the thighs are braced, and the body assumes a spiral bend over the right side beyond the base line, so the eyes can see the heel of the left foot. Weight is nearly divided, but stronger on left foot. The intent here shows strength. (For base line see Laws of Gravity.)

STRAIGHT (knee bend) ATTITUDES.

Assume these Attitudes in the order given until familiar with them.

No. 1.
STRAIGHT (knee bend) ATTITUDE.
Assume a Straight Attitude and bend both knees. Arms in 3d position.

No. 2.
STRAIGHT (right knee bend) ATTITUDE.
Assume a Straight (right side) Attitude and bend right knee; body front view. Arms in 3d position.

No. 3.
STRAIGHT (left knee bend) ATTITUDE.
Assume a Straight (left side) Attitude and bend left knee; body front view. Arms in 3d position.

No. 4.
STRAIGHT (forward knee bend) ATTITUDE.
Assume a Straight (forward) Attitude and bend forward knee; body front view. Arms in 3d position.

No. 5.
STRAIGHT (back knee bend) ATTITUDE.
Assume a Straight (back) Attitude and bend back knee; body front view. Arms in 3d position.

Execute Nos. 4 and 5 at a ⅞ front view.

PRACTICE.

Physical Corrections, sinking and rising (VII. No. 2, page 25) with Attitude No. 1.

Exercise No. 3. The Wave Lines (Wave Series page 76) with Attitudes Nos. 2 and 3.

Exercise No. 4. The Warrior (Spiral Series page 88) with Attitudes Nos. 4 and 5.

Do the above continuous, blend from one Attitude to the other; using at same time the proper action and Arm Movements. Gavotte music.

EXPLANATION.

In the study of Physical Corrections (page 17,) will be found the study of the Knee (sinking and rising.) The bend of the knee with Attitudes is used to obtain the Spiral Motion of the body. In stationary positions the knee bend adds strength to the Attitude.

The leg is the organ of locomotion and seat of physical strength (see the Elements of Man.) The spiral movements start from the bending of the knee and relate to Force; therefore the knee bends give the Attitudes force and strength.

In the study of the Triune of Attitudes, the Straight Attitudes relate to the mental, the Arch Attitudes to the emotional elements (being curved); the knee bend added to them completes the trinity of elements—which gives them physical strength or power.

NOTE.—In the Study of Attitudes the basic law of the trinity has been followed and is shown from different standpoints. The division of seven has not been followed in this instance, for reasons of imparting the study as easily as possible; but for the critical student the proper divisions are given. In Straight Attitudes 5 are shown, Nos. 4 and 5 are again executed in oblique formation, completing the 7 Straight Attitudes; Nos. 1, 2 and 4 being the trinity. In Arch Attitudes 4 are shown, Nos. 1 and 2 of Spinal Column Attitudes, and No. 3 of Beyond Base Line completing the 7. The Triune of Arch (knee bend) Attitudes and Spinal Column Attitudes are illustrated. The Mercury Balance Attitude and Nos. 1 and 2 of Beyond the Base Line, also make a trinity, as the same law of Gravity governs them.

ARCH (knee bend) ATTITUDES.

Assume these Attitudes in the order given until familiar with them.

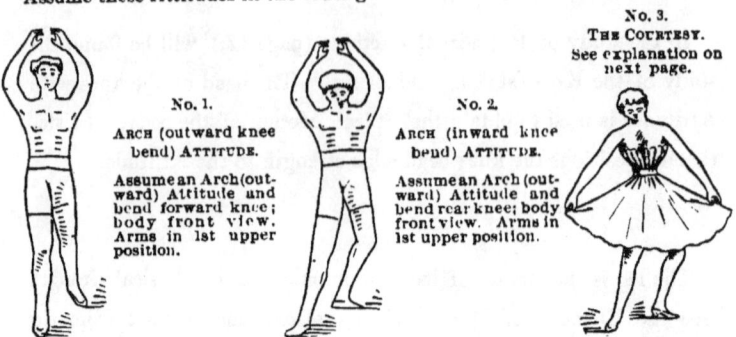

No. 1.
ARCH (outward knee bend) ATTITUDE.
Assume an Arch (outward) Attitude and bend forward knee; body front view. Arms in 1st upper position.

No. 2.
ARCH (inward knee bend) ATTITUDE.
Assume an Arch (outward) Attitude and bend rear knee; body front view. Arms in 1st upper position.

No. 3.
THE COURTESY.
See explanation on next page.

PRACTICE. { Exercise No. 3 (Spiral Series page 88) with Attitudes Nos. 1 and 2.
Exercise No. 7 (Spiral Series page 92) with Attitudes Nos. 1 and 2.

SPINAL COLUMN (bend) ATTITUDES.

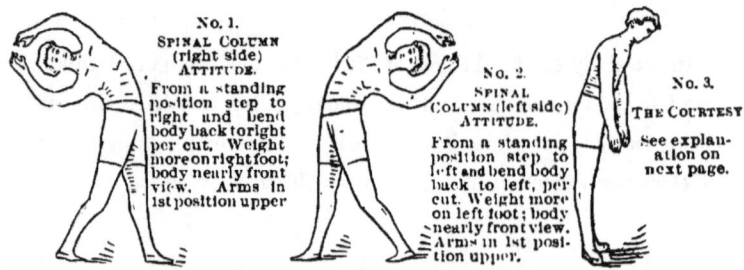

No. 1.
SPINAL COLUMN (right side) ATTITUDE.
From a standing position step to right and bend body back to right per cut. Weight more on right foot; body nearly front view. Arms in 1st position upper.

No. 2.
SPINAL COLUMN (left side) ATTITUDE.
From a standing position step to left and bend body back to left, per cut. Weight more on left foot; body nearly front view. Arms in 1st position upper.

No. 3.
THE COURTESY.
See explanation on next page.

PRACTICE. { Physical Corrections (III. No. 2 p. 35) with Nos. 1 and 2.
Physical Corrections (II. No 2 p. 35) with Nos. 1 and 2 and *Pirouette*.

BEYOND THE BASE LINE ATTITUDES.

No. 1.
SIDE VIEW (right side) ATTITUDE.
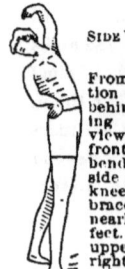
From standing position circle left foot behind right presenting the right side view of the body as a front view; again bend body over right side as per cut; left knee bent. Thighs braced and weight nearly equal on both feet. Left arm in 1st upper position and right as a counter poise.

No. 2.
SIDE VIEW (left side) ATTITUDE.
From standing position circle right foot behind left presenting the left side view of the body as a front view; again bend over left side as per cut; right knee bent. Thighs braced and weight nearly equal on both feet. Right arm in 1st upper position and left as a counter poise.

No. 3.
THE BACK BEND.

From standing position step back with right foot; brace yourself by bending both knees and bend backward. Use the bend of the neck and head to complete the arching.

PRACTICE. { Physical Cor. (III. No. 3 p. 35) with Attitudes Nos. 1 and 2.
No. 3. Slowly ½ pirouette, bend back and recover. Slow gavotte time

EXPLANATION.

The Courtesy.

The Courtesy (for ladies) standing position; circle right foot behind left, assume an Arch (inward) Attitude, sinking well as per cut, fingers draping dress; observe toe position and poise of head. To recover, rise to an erect Attitude and return the back foot with a circle movement to place. Repeat with left foot. Execute the Courtesy at ¾ view front. Slow gavotte music.

The Bow.

The Bow (for gentlemen) standing position; hands at side, step to right and place feet together, same time face right; now bend forward, allow the hands to droop loosely in front, as per cut. To recover, assume erect position, hands fall to side and at the same time body face front. Now repeat to left. Slow gavotte music.

Arch (knee bend) Attitudes.

The peculiarities of the Arch (knee bend) Attitudes are worth knowing well, as in most cases they denote some individuality or impersonation. A technical analysis of these Attitudes shows that the Arch bend consists of part of a circle, which requires flexibility and the knee bending adding character or force; further these Attitudes always have a mental meaning, giving to them an element of the three forces.

Spinal Column Bends.

The Laws of Gravity (page 16) relating to counter poise apply to these bendings. It is also well to read Study of Spinal Column and laws relating to same (Physical Corrections page 38.) In Bendings the knee, hip, neck and head complete the arching.

Beyond the Base Line.

These Attitudes are Spinal Column Bends, adding the Thigh Brace (page 24,) which allows the body to bend beyond the base line, and calls for the Laws of Gravity relating to balancing the body beyond the base line.

KNEELING ATTITUDES.

Assume these Attitudes in the order given until familiar with them.

No. 1.
KNEELING (front) ATTITUDE.
Assume a Straight (forward knee bend) Attitude and kneel down. Arms in 3d position.

No. 2.
KNEELING (right) ATTITUDE.
Assume a Straight (right knee bend) Attitude and kneel down; face and body front view. Arms in 3d position.

No. 3.
KNEELING (left) ATTITUDE.
Assume a Straight (left knee bend) Attitude and kneel down; face and body front view. Arms in 3d position.

RECLINING ATTITUDES.

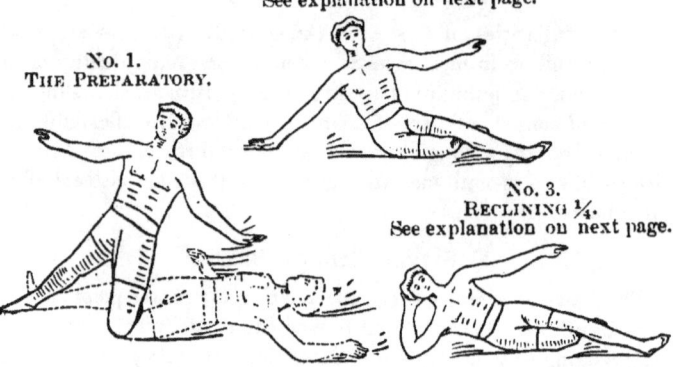

No. 1. THE PREPARATORY.

No. 2. RECLINING ¼. See explanation on next page.

No. 3. RECLINING ½. See explanation on next page.

LYING ATTITUDES.

No. 1. LYING ON SIDE.
No. 2. LYING ON BACK.
No. 3. LYING ON FACE.

See explanation on next page.

PRACTICE.

Arm Movements No. 1. The Harmonic Three (Circle Series page 70) with Kneeling Attitude No. 1.

Arm Movements No. 2. The Complete Circle (Circle Series page 70) with Kneeling Attitude No. 2.

Arm Movements No. 5. The Complete Circle (Circle Series page 70) with Kneeling Attitude No. 3.

EXPLANATION.

KNEELING ATTITUDES.

The Kneel (front) Attitude may be executed at a ¾ angle front view; also right side or left side view or leaning backward. The right or left Kneel Attitudes are similar to the Straight (right or left side knee bend) Attitude, only kneeling; the face and body front view. They will admit of all changes of the Straight or Arch Attitudes; starting the base line from the knee on floor to the head. Arm Movements are also executed with these Kneeling Attitudes, all laws of Poise, Opposition and Harmony being observed.

RECLINING ATTITUDES.

From a Kneeling Attitude, start to recline as per cut No. 1, to a half Reclining Attitude, allowing the hip and waist to rest upon the floor, and the palm of the under hand acting as a support to that part of the body still up as per cut No. 2. In the next Attitude, "Reclining ¼," the hip, the side of trunk, and arm to the elbow rest upon the floor, with hand supporting the head, and the top arm free as per cut No. 3.

LYING ATTITUDES.

In lying down on the side, the under leg should be crossed and the top leg straight; this law applies to the body when down after falling. The body is now prepared to arise to the Reclining Attitudes, then to kneeling, then to rising; so perfect a state may this practice become that not a single garment of the wearer will be disturbed in falling or rising. Lying on back or face would only require turning over to a side position thence to rising as just stated.

POISE.

1. MERCURY BALANCE POISE.
2. AN ARABESQUE POISE.
3. POISE OF DIFFERENT PARTS OF THE BODY.

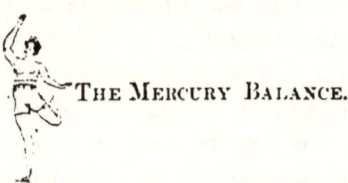
THE MERCURY BALANCE.

In the Mercury Balance, the body is poised or balanced on one foot, and well illustrates the poise of the body. This poise can easily be accomplished and is fully explained in the Study of Gravity, which includes the Mercury Balance exercise and its derivations.

AN ARABESQUE.

The Arabesque is an Arch (right or left knee bend) Attitude, using the complete arm circle movement with a spiral movement of the body; the spiral motion occurs at the change of Attitude from right to left. The change here shifts the control of gravity through its various centers in the body.

POISE OF DIFFERENT PARTS OF THE BODY.

To Poise any part of the body is to give it the mental prominence for the time being and balance such part by creating the proper counter poise, which in reality completes the poise or balancing. The head, arms and legs will permit of this liberty. Derivations from these principles are such as the nose, chin, shoulders, elbows, hands, fingers, toes, heel, knee and hip.

FORCE.

Force is an unseen power quite unintelligible. We can not imagine it except through the instrumentality of matter, which only postpones our difficulty so far as conceivability is concerned. Strange as it may seem we are equally unable to grasp its mode of exercise. Where there is life, Force is there in a greater or less degree. We say that as far as our consciousness and the agencies producing the various manifestations related thereto, that between the physical forces themselves and the sensations, there exists a correlation which stands in direct correlation with physical forces existing externally. These, of course are analytical truths, none of which will constitute that synthesis of thought which is necessary to be an exposition of the synthesis of things.

FORCE.

In the Study of the Grace of Man, the intention of this work is not to expound a treatise on what Force is composed of or where it may originate, or what law governs its action or exercise. Its *existence* is all we are concerned with, together with its further development.

Force is generated from the different elements in man, and may emanate from the Soul, the Mind, or the Body. Where there is life, there will be found Force in a greater or less degree. It is the outward expression of such life elements within, and is in evidence according to the development of the organism.

By perfecting the physical body the Physical Power is developed. By developing the mind you increase the Mental Force. The Spiritual Force is developed when the soul is given freedom of expression. The soul expression through the mind is Personal Magnetism. How to develope these Forces is given fully in the pages that follow and is divided into seven subjects, as follows:

FORCE IS POWER

And consists of
1. Reserve Force —Mental —Gravity.
2. Spiritual Force—Emotional—Flexibility.
3. Physical Power—Physical —Strength.

DERIVATIONS.

4. Will Power.
5. Personal Magnetism.
6. Propelling Force—Walking.
7. Hypnotism.

Reserve Power.

Reserve Power is the control of Forces which have been cultivated by the different exercises throughout this work (sometimes they are inherent.) Power should never be entirely exhausted; some of it should be held in reserve.

Mental Control.

The necessity of Reserve Force is ignored by the extremist, whereas the Mental Control should always be in a condition to have a supply in store; this keeps up the Emotional Force which draws upon the physical to a great extent.

The Law.

The law of equality is plain in the case of Reserve Force. If you exhaust one Force, you do it at the expense of another. Thus it is that some persons are all mental and emotional, or all physical; in fact various are the combinations of such anomalies, so to speak.

EXERCISES

FOR THE CULTIVATION OF RESERVE POWER.

From the various studies in this work.

I. GRAVITY. The Pirouette Exercise, page 10.
FLEXIBILITY. The Relaxation Exercise, page 100.
FORCE. The Entire Study, page 121.

II. PHYSICAL CORRECTIONS SERIES I. II. III.
Thigh Brace Exercise page 20.
Waist Spanish Circle, page 32.
Draw Back Bends, page 34.
The Groins for Breathing, page 36.
Fingers (separate control), page 44.
Hand (beauty position), page 44.
Head (side to side) Exercise, page 44.

III. STUDY OF CURVED LINES.
The Repentant (No. 6 Circle Series,) page 72.
The Inspiration (No. 7 Circle Series,) page 72.
The Upward Advance (No. 4 Wave Series,) page 78.
The Gateway of Heaven (No. 7 Wave Series,) page 80.
The Warrior (No. 4 Spiral Series,) page 88.
An Allegory (No. 6 Spiral Series,) page 90.
The Arabesques (No. 7 Spiral Series,) page 92.

IV. STUDIES OF GRACE.
All studies should show latent intensity of Reserve Force for their expression.

V. STUDY OF ATTITUDES.
Beyond the Base Line Attitude, page 116. Poise, page 120.

VI. STUDY OF HARMONY AND OPPOSITION. (Page 135.)
This entire study requires some form of Reserve Force for the proper execution in transmitting the intention of the Attitude assumed.

VII. THE MIND.
Read Aphorisms for the development of the Mind.

Spiritual Force.

Spiritual Force is that indefinable something which emanates from a strong free soul, that has been illuminated by the divine spark. The stronger the soul the more of the divine spark will it be able to absorb and the greater will be the soul's radiance.

Duality.

No known law governs the soul's action of Force. It comes from that high source of which this work merely hints. Its duality is shown in the Physical (Physical Power,) "what is in the spiritual is also in the flesh;" what it is here, we can only perceive through Personal Magnetism, and develope it by correct living.

Manifestation.

Spiritual Force is the love essence of the soul and is only made manifest after the body has been purged of all carnality and the soul has become its master.

Aphorisms

To Cultivate Spiritual Force.

1. Abstinence. Temperance.

2. Self denial.

3. Follow the laws of nature always.

4. Observe the laws of health.

5. Self inflicted penance.

6. Soul meditation.

7. Prayer.

Physical Power.

Physical Power is strength and according to the Anatomy of the Elements of Man this power is located in the extremities. The arms are for protection and the legs for locomotion, not forgetting the trunk from which these branches grow.

Walking.

Perfect walking embraces the perfection of the Physical Corrections of the body with a complete harmony of the study of Gravity, Flexibility and Strength. In walking there should be gravity buoyancy of the body, errect carriage and proper setting of the head. The laws relating to toe, heel and ankle or foot must be carefully studied. For Walking Exercises, see page 129.

Strength of Muscles.

The muscles in the arms, legs and body should be flexible; the strength of steel with the flexibility of leather.

WALKING EXERCISES.

No. 1. (THE BASE.)
Practice Toe Position, page 20. Practice Ankle Twist, page 20. Now walk showing these corrections. March time.

No. 2.
Practice Buoyancy, page 10; same time execute Elbow Twist, page 44. Now walk showing all of No. 1, with Buoyancy added. Buoyancy is for the natural adjustment of the muscles. March time.

No. 3.
Assume Lock Exercise, page 32; same time do Toe Position, page 20; next, same time do Buoyancy, page 10. Now walk with hands locked, showing previous exercises. Care should be taken to have erect carriage. To correct any imperfection lock hands higher. March time.
NOTE—Chin in, chest out.

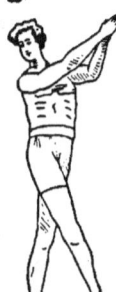

No. 4.
Clapping hands at 3d upper front position, then swinging arms to 3d lower behind, (as per cut) Do this and walk, observing all previous rules. Waltz time, then March time.

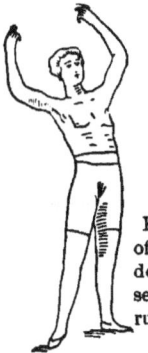

No. 5.
Practice overhead sway of arms, page 46. Now do this and walk, observing all previous rules. Waltz time.

No. 6.
Place arms in 3d or balance position and walk. This exercise is for strengthening the muscles under the arms, giving a freedom of extremities. March time.

No. 7.
Practice Running Exercise, page 14, breathe through nostrils, observe head position, elbow crooked and hands closed. Now run (double quick) keep time. Polka time.

WILL POWER.

"He that controlleth his temper is greater," etc., is well known to every person and this can only be accomplished by controlling the will. The whole foundation of the universe is based upon Will. If all the prominent religions are carefully sifted they will be found to rest upon will. Our will is but a reflection of the Supreme Will, therefore its limitations or possibilities are unknown to us. Having developed the mind the next essential step is to exercise constant control of that will.

HABIT.

Lack of Will Power is a weakness and is a bad habit; it can be corrected by gravity and strength of mind and following the aphorisms given on next page.

TEMPER.

Will Power and stubbornness are often confounded; but there is a vast difference. Stubbornness will cling to a subject or point against all reason, or possible gain thereby, showing a decided weakness of intellect. Persons who lose their temper lack Will Power to control it.

Aphorisms

To Cultivate Will Power.

1. Reserve the Mental Powers.

2. Maintain Gravity of Mind.

3. Control of Temper.

4. Purge yourself of Laziness.

5. Do not let any one element predominate. Extreme enthusiasm knows no law of harmonic reason.

6. Steadfastness to Principal and Firmness of Purpose.

7. Purity of Thought.

Personal Magnetism.

Personal Magnetism is that force that is the glow of the soul, as the flower has its perfume, the sun its sunbeam, both of which are rays of force. Personal Magnetism is only in evidence when the soul has dropped its fetters or blossomed. It is there at all times in a positive or negative degree.

The Law.

The law is simple and the proof is to try the law. Perfect the body; express the intellect of the mind through and by movements of the body, until the perfection and intensifying of these expressions calls for a deeper force which comes from the soul. This deeper force of soul and mind is Personal Magnetism. The practice of intensifying the proper mental expressions and emotions, establishes the channel that reaches the soul and Personal Magnetism is developed. The soul is thus given freedom of expression through Personal Magnetism.

Exercises.

Personal Magnetism relates to the soul, therefore all exercises to cultivate it, must be devoid of the grosser elements. Maintain or practice the mental, the spiritual and the spiral movements. The grosser elements will not contribute to the cultivation of Personal Magnetism. See the Study of the Soul and the Anatomy of the Elements of Man.

EXERCISES

For the Cultivation of Personal Magnetism.

From the various studies in this work.

I. GRAVITY. Buoyancy Exercise, page 10.
FLEXIBILITY. The Entire Study, page 94.
FORCE. Reserve Power, Spiritual Force and Will, page 121.

II. PHYSICAL CORRECTIONS, SERIES I. II. III.
Waist Spanish Circle, page 32.
Abdomen and Stomach Circle, page 32.
Wrist, (figure eight,) page 48.
Hand, the discharge, page 48.
The Spiral of the Hand, page 48.
Head and Vertebra Exercise, page 48.
Head, Circle Movement, page 48.

III. THE STUDY OF CURVED LINES.
Transmission of Power, (No. 5 Circle Series, page 72.)
The Emotion of Love, (No. 5 Wave Series, page 78.)
In a Garden Fair, (No. 6 Wave Series, page 80.)
The Spiral Motion, (No. 1 Spiral Series, page 86.)
The Call to God, (No. 2 Spiral Series, page 86.)
Take my Soul in Thy Keeping, (No. 3 Spiral Series, p. 88.)
An Allegory, (No. 6 Spiral Series, page 90.)

IV. STUDIES OF GRACE.
All studies show a flow of Personal Magnetism governed as it were by Reserve Power.

V. STUDY OF ATTITUDES.
The Knee Bend Attitudes and exercises relating to them, page 118; also Arabesque Attitudes, page 120.

VI. STUDY OF HARMONY AND OPPOSITION.
The entire study of Harmony must be applied to all movements for the proper development of Personal Magnetism, page 135.

VII. The study of the Soul, the Mind, the study of Ideality and Inspiration, complete the study of Personal Magnetism.

HYPNOTISM.

Hypnotism is physical magnetism and is demonstrated through the control of will. Hypnotism has no direct bearing on this work, it is merely mentioned because it is one of the manifestations of the Force of Will.

POWER.

Hypnotism proves in this work that an unseen force has a power, which is an acknowledgment that all the Forces mentioned have a power of their respective kinds.

REASON.

In a state of trance the consciousness is entirely mastered; the mind may even forget the existence of the physical body. Man exists as an individual just so long as he is in possession of that divine reason he is originally endowed with.—A reason which is not an attribute of the human form, but a function of the divine spirit which illuminates it.

Poise, Opposition and Harmony.

POISE, OPPOSITION AND HARMONY.

POISE—Gravity—Mental.
OPPOSITION—Force—Emotional.
HARMONY—Flexibility—Physical.

There is a Harmony in every attitude and movement in this work. In all things, God, Man and the Universe.

Duality implies Opposition. The Harmony of Duality is where the third element is added which constitutes a Trinity and completes a harmony of the whole. Harmony is in this way established; for instance:—The Trinity of the Body, the Mind and the Soul; thus at a glance in any study you should be able to find the harmony. To take one element from a Trinity breaks the harmony.

The Grace of God is the harmony in the Trinity of the Father, Son and Holy Ghost. The Grace of Man is the harmony in the Trinity of the Body, the Mind and the Soul. This (Triune) Harmony will be found throughout all the studies in this work.

To Poise the body is to balance the body or parts of it. (See Study of Attitudes, page 120.)

Opposition is to place in counter poise a part of the body to complete a balance which creates a Poise.

Harmony of two parts of the body in different locations is to create a sympathy between them, or a leaning toward each other; which is equalizing them to a balance and with an opposition, which is a counter poise (or balance,) constitute a harmony of the whole.

A Harmony of parts in a balance position with an Opposition (creating a counter poise) is a complete Poise.

EXEMPLIFICATION.

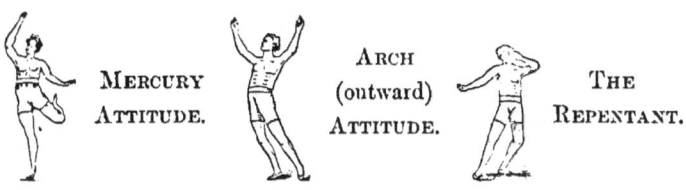

Mercury Attitude. Arch (outward) Attitude. The Repentant.

Mercury Balance Attitude.

The body is balanced on one foot. The Harmony is established by the inclination of the head and acting foot (left) creating a balance. The waist is opposed, (convex at left side) which creates the counter poise or counter balance, making a complete poise. The head is inclined from the high (right) hand; the left in position acting as a counter balance to right hand. The line of Gravity is observed from neck to center of heel of standing foot.

Arch (outward) Attitude ¾ view front.

The Harmony is established by the inclination of the head backward to a balance with the acting foot, which is the rear foot. The opposition of the waist occurs in the back of the waist (back convex.)

The "Repentant" Attitude.

The pointed foot in front (right) and the head forward and to right, create a Harmony of two parts. The opposition of the trunk is front of waist (near right side); the head is inclined from the high hand.

HARMONY.

The Harmony of Gravity is a Perfect Balance.

Elements
- Mental—Rhythm, Reserve Power, Control of Force.
- Emotional—Bouyancy, Counter Poise, Blendings.
- Physical—Balance, Regularity of Motion, Equilibrium.

The Harmony of Flexibility is Ease of Motion.

Elements
- Mental—Perception, Intent, Directs Force.
- Emotional—Action, Force, Relaxation.
- Physical—Ease of Motion, Control of Muscles, Individual Control.

The Harmony of Force is Control of Power.

Elements
- Mental—Reserve Force, Gravity of Mind, Will Power.
- Emotional—Spiritual Force, Personal Magnetism, Hypnotism.
- Physical—Physical Strength, Propelling Power, Distribution of Force.

In the Study of the Elements of Man,

The Trinity of God by its completeness constitutes Harmony in its fullest sense. This applies also to the Trinity of man. Neither would be complete if one factor were missing. This completeness constitutes Harmony, and here we have the foundation of Harmony.

The Harmony of Curved Lines is Perfect Grace.

Mental—Circle Lines.
Emotional—Wave Lines.
Physical—Spiral Movements.

The Harmony of Grace is

The Expression of the Soul,
Through the Mind,
By Motion of the Body.

Laws of Opposition and Harmony Showing the Harmony of the Whole.

Laws of Opposition and Harmony showing the Harmony of the Whole.

Applied to each part of the body.

SERIES I.

The Toes, Heel and Ankle.

The toes and heel should always show a natural law of opposition, that is toes in one direction and heel in an opposite. The use of the ankle completes the Harmony of the proper poise of the foot.

The Heel, Hip and Knee.

The heel, knee joint and hip muscles are related and make a uniform whole, which is a Harmony. The heel and hip the Harmony with the knee as a counter poise.

The Knee, Calf of Leg and Thigh.

There is a natural law of Harmony of the knee, calf of leg and the thigh. The latter two parts should close evenly at the bending of the knee to create a good balance, constituting the Harmony. In Bouyancy, the opposition is the counter poise of the knee itself.

BOUYANCY—THE KNEE, SHOULDERS AND HEAD.

The Law of Harmony in Bouyancy is a balance of the knee action with the shoulders. The shoulders are the pedestal for the head; the head always directs the Harmony of a poise. The opposition (counter poise) in this case is the trunk at the waist line, observing the centre of gravity of the entire body.

THE LEGS, THE TRUNK, THE HEAD.

The natural law of the legs are their opposition movement in propelling force or walking, which balances the body; the trunk (especially the chest,) the head and arms make a complete weight and is a counter poise which the legs carry. The balancing of the whole is the Harmony.

THE ACTING FOOT, THE HEAD AND THE WAIST.

The acting foot is usually in Harmony with the head, and the waist is opposed. The waist is opposed (and concave) at the point where this Harmony, " head and acting foot" is established, more especially in attitudes of poise on one foot. The waist line meaning around the body and this counter poise may occur at any part of it.

THE HIP, THE KNEE AND THE ANKLES.

The Harmony of ankle, knee and hip muscles should be well studied, as they are the hinges of the leg. In Arch Attitudes this Harmony is established in its most pronounced sense.

SERIES II.

The Waist, Control of Centre and Freedom of Extremities.

The waist is the supposed centre of the control of gravity. This control of gravity, with a freedom of the extremities form a complete Harmony. This center is subject to an opposition poise or Harmony to retain the centre of gravity of the whole.

The Trunk, the Chest and Abdomen and Stomach.

The Harmony of the trunk is the carriage of the chest, abdomen and stomach with an opposition of the arms which completes its proper carriage.

The Spinal Column or Back, Shoulder Blades and Chest.

A perfect Harmony is created by strength of spinal column with an opposition balance of the shoulder blades; this gives the trunk (or chest) its proper carriage. This Harmony creates a physical support or Harmony of strength.

Abdomen, Stomach and Groins.

The abdomen, stomach and groins form a complete Harmony; either part not being in its proper position would destroy the Harmony of the waist line.

THE CHEST, THE CHIN, THE SHOULDER BLADES.

 The chest is the mentality of the trunk and relates therefore to the head. The relationship is first to the shoulder blades as a support to the chest, and second the chin drawn in, in opposition to the chest; forming a Harmony of shoulder blades and chin; viz: chest out and chin drawn in, with the shoulder blades, constitute a complete Harmony.

BREATHING—LUNGS, DIAPHRAM AND GROINS.

 The lungs' action, the storage at diaphram and the added control of muscles of the groins constitute the Harmony of proper breathing.

THE NECK, THE HEAD AND THE SHOULDERS.

 The muscles of the neck is subservient to the head muscles and those of the shoulders, viz: the head rests upon the shoulders connected by the neck, and the Harmony is a natural one. The neck should denote flexibility.

SERIES III.

THE ELBOW, THE SHOULDER AND THE HAND.

 The elbow is the natural opposition of the shoulder and hand. The perfect blending of shoulder joint, elbow, wrist and fingers should be given great care. A correct Harmony of curves, wave lines and spiral lines are executed with the arm. The elbow being the emotional part of the arm, such force is always apparent at the elbow and should express this element in a complete Harmony of the whole arm.

THE WRIST, THE ELBOW JOINT AND THE SHOULDER JOINT.

The wrist will bear the same Harmony as that of the ankle and neck. Its character is of a more flexible nature, having a more intellectual duty to perform than either of the two mentioned. The Harmony of transmitting strength and curve is from the shoulder to the elbow, to the wrist; and the even blending of these parts also constitutes a Harmony.

THE HAND, THE FINGERS AND THE HEAD.

 In all expressions there should be a perfect Harmony of the hand and head. The hand is the physical agent of the mind and its third element is intelligence, which completes one of the most subtle harmonies. The physical Harmony of the hand is clearly shown in the beauty position; the fingers and lower cushion being in balance, the counter poise of the wrist completing the Harmony of the whole.

THE ARM, THE LEGS AND THE HEAD.

The Harmony of the arms is their natural law of opposition swing, which creates a balance. This swing is in opposition to the natural opposition of the legs. From these natural laws, the deduction is easily made that the right foot and left arm are always poised in opposition to create a balance, and the same rule applies to left foot and right arm, in any attitude. Opposition movements of the arms are done simultaneously. Parallel movements follow each other. The arm is the channel of expression of the mind to the hand. This last Harmony is of a high order and relates to Mental Force. The same Harmony is found in the Emotional and Physical Forces when applied to the arm.

THE SHOULDERS, THE CHEST AND THE HEAD.

The shoulders should balance each other, and the chest in opposition completes the physical Harmony. The thought of comfort and ease transmitted to the feeling of the shoulders indicates their proper adjustment. The shoulders reflect various intents of the mind on this basis.

THE FINGERS—THE THUMB, THE INDEX AND LITTLE FINGERS.

The Harmony of the fingers is in their expression of curve, flexibility and force. The harmony of the two middle fingers and thumb and opposition of index and little fingers form a physical Harmony of beauty.

NOTE—The entire hand presents the same form of harmony as that of the body as a whole; four distinct parts, viz: Thumb, palm, index and little fingers, and middle fingers, representing head, trunk, arms and legs. These four parts harmonizing under the head of three.

THE HEAD.

The head is the fountain of Harmony of the body and there is no complete Harmony of any parts of the body without some expression of the head. Herein are found the natural laws of Harmony. The agents of the machinery within are located in the different parts of the head. The head is always in Harmony and forms a balance with the acting foot; the trunk is opposed. The head is always opposed to the highest hand or hands. The poise of the head first indicates and at last finishes the Harmony of gravity in assuming any attitude or position.

CORROLLARY.

The extremities (arms and legs) are opposed by a natural law, which constitutes a Harmony.

The trunk is opposed (a counter poise) which completes a poise.

The head is opposed to either a leg or arm, and at the same time is in harmony with the opposite leg or arm, and the head inclines to start and finish the gravity of a poise.

Analysis of Attitudes.

(Showing Harmony and Opposition.)

STRAIGHT ATTITUDES—(For Illustrations, see page 108.)

Straight (standing) Attitude.
The Harmony is the gravity of the whole body.

Straight (right) Attitude.
Harmony—head and left foot, left waist opposed.
Gravity line from neck to right heel.

Straight (left) Attitude.
Harmony—head and right foot, right waist opposed.
Gravity line from neck to left heel.

Straight (forward) Attitude.
Harmony—head (back) and back foot, back waist opposed.
Gravity line neck to right heel.

Straight (back) Attitude.
Harmony—head and right foot, front waist opposed.
Gravity line neck to left heel.

The Mental Harmony should be equilibrium or gravity transmitted to the mental centers—namely from the head to the chest and hands.

ARCH ATTITUDES—(For Illustrations, see page 110.)

ARCH (right) ATTITUDE.
Harmony—head and left foot, left waist opposed.
Head opposed to right hand. Gravity line neck to right heel.

ARCH (left) ATTITUDE.
Harmony—head and right foot, right waist opposed.
Head opposed to left hand. Gravity line neck to left heel.

ARCH (outward) ATTITUDE.
Harmony—head (back) and left foot, back waist opposed.
Gravity line (perpendicular) neck to right heel.

ARCH (inward) ATTITUDE.
Harmony—head (forward) and right foot, front waist opposed.
Gravity line (perpendicular) neck to left heel.

The Mental Harmony is flexibility of muscles transmitted to the muscles comprising the arch. Also the Harmony of gravity transmitted to the mental centers perfecting the gravity.

BENDING ATTITUDES—(See page 112.)

STRAIGHT (forward knee bend,) ATTITUDE.

The mental harmony is first established—usually heroic. The physical harmony is the same as Straight (forward) Attitude, the bend of forward knee giving the Attitude force. Arms are in counter poise; the harmony is of head and acting (rear) foot; head in opposition to high hand; gravity line is from standing front foot to neck; all form a correct harmony.

ARCH (inward knee bend) ATTITUDE.

The mental harmony is first established—usually sorrow, etc. The physical harmony is the same as Arch (inward) Attitude, the bend of rear knee giving the Attitude force. Arms are in counter poise; the harmony is of head and acting front foot; head inclined from high hand; front waist opposed; gravity line from standing rear foot to neck (perpendicular); all form a correct harmony.

SPINAL COLUMN (side view) ATTITUDE.

The mental harmony is first established—gravity beyond base. The left foot is placed behind the base, (now standing side view). The head and right foot are placed in harmony, (the head away from high hand) and gravity line established head over base; now bend to right beyond base line, (follow the laws of gravity), the head leading and finishing the poise intended.

THE COURTESY.

The mental harmony is dignity, (mental for Bow is obedience), the physical harmony is the same as Arch (inward) Attitude, the bend of rear knee giving the Attitude expression. Arms are balanced; the harmony is of head, (inclined to right and forward) and front foot; front of waist opposed. Every movement required for the execution of this Courtesy will bear a technical analysis as prescribed in the Art of Dancing, viz: "*Assemblé, Demi-Ronde Jombe. Balancé, Sorté. Demi-Ronde Jombe Assemblé.*"

SPINAL COLUMN (right side) ATTITUDE.

The mental harmony is strength of poise and gravity beyond the base line; face front, step to side with right foot, forming a solid base; bend over to right, follow laws of gravity, using the head, hands and bend of knee and arm to complete the harmony of the arch or circle.

THE ARABESQUES.

The mental harmony is grace of motion. The Arch is of the right side. The physical harmony is of the head and left foot; left side waist opposed; head inclined from high hand; arms are counter poised. In the execution of this Attitude from right to left the harmonies of Spinal Column Attitudes, the Arch Attitudes and Straight Attitudes must be observed, as the body passes through them, to reach an Arabesque.

THE BACK BEND.

The mental harmony is strength of poise and gravity beyond the base line. About face, step back with right foot forming a solid base, bend backward and follow the laws of gravity, using the head, neck, knees and hips to complete the arching, creating a harmony of the subjective centres of gravity.

In the study of the Triune of Attitudes, the Straight Attitudes relate to the mental, the Arch Attitudes to the emotional elements (being curved); the Knee Bend added to either completes the trinity of elements—which gives them physical strength or force.

The Mind.
The Inspiration Theory.
Ideality.

The Mind.

The Mind is that force or essence of forces that creates intelligence and is the fruit of that instrument known as the brain. Here is the seat of intelligence. This brain is the receiving office and dispatching depot to the different mental centres of the body from which this intelligence is then distributed to the various parts of the body, which are the subjective centres, *i. e.* the hand, the chest, etc. The mind may be developed, in fact it is nearly a natural law. The muscles of the body may be developed, and this again proves that the soul may be developed.

THE MIND.

Its divisions are 1. Reason, Conception, Perception.

2. Receives Inspiration, Inspires, Ideality.

3. Imparts Force, Intelligence and Magnetism.

DERIVATIONS.

4. Conception and Ideality.

5. The Inspiration Theory.

6. The Intelligence.

7. Its Relation to the Soul.

Axioms

For Developing the Mind.

1. Reason.

2. Be patient and observe closely.

3. Control your temper.

4. Study mathematics.

5. Purge the mind of impure ideas.

6. Think, study and conceive.

7. Study the Soul.

INSPIRATION.

An Inspiration is a message from the Soul to the Mind. It may be a thought, a conception or a force.

In the study of Personal Magnetism in this work mention is made of its being the glow or force of the Soul. When this Personal Magnetism carries with it an intelligence, it is then Inspired Personal Magnetism or Inspiration.

Inspired Personal Magnetism is Personal Magnetism in its most developed form. If we possess Personal Magnetism we will soon be permeated with Inspiration of the Soul; and be able to inspire ourselves and carry others with these inspirations or intelligences. Again, if Personal Magnetism can be imparted, we can transmit to, or inspire others and they will feel this force.

IDEALITY.

The beautiful study of Ideality is developed in the mind. When the mind is touched by an inspiration, thoughts are clothed in golden garb and perfection is king. Ideality leads to purity of thought and elevation of the mind for the development of the beautiful.

Ideality cannot be fostered by any of the grosser elements, as it is the first sign or touch of the spiritual force of the soul.

Thus does the Soul in time register its various forces in the mind, for its deliverance to freedom.

The registering of the various forces in the mind creates intelligence. A force is thus established, which is for the purpose of imparting such knowledge through the various instruments of the body, viz: the mouth to speak, the eyes to express, etc. This intelligence directs all forces.

The Study of the Soul.

The Soul.

GOD
The Father, Son and Holy Ghost.
The Grace of God.

Love.   Wisdom.   Power.

THE SOUL

Has an Intelligence Force.
The Life Force within us.
A Spiritual Force.

This Trinity in One is the Soul.

Intelligence is Wisdom.
Life Force is Power.
Spiritual Force is Love.

Thus do we resemble the Great Creator only on a lesser scale.

The Soul is the reflection or "*part of*" the all prevailing "God."

The word reflection here represents a Force. The flowers have their odor, the sun its rays, the earth its atmosphere, the Soul its force, and God, His reflection.

GOD.

Wisdom. **Love.** **Power.**

THE SOUL.

Intelligence Force. **Spiritual Force.** **Life Force.**

THE BODY.

Mental Element. **Emotional Element.**
Physical Element.

The harmony of the Soul is the forces mentioned; without either it would not be a soul.

RELATION.

The **Intelligence Force** is related to the Mind.
The **Spiritual Force** is related to the Heart.
The **Life Force** is the inner atmosphere of our Body.

The Trinity of these three forces forms a magnetism similar to that given in the Study of the Mind. This is Personal Magnetism. It should rightfully be termed Soul Magnetism.

TO CULTIVATE THE SOUL.

To cultivate the Soul, so do as you cultivate the plant. The soil (the body) is carefully tilled. The flower grows, and its careful training shows in its beautiful physical formation. Its perfume is an offer of unbounded joy for its freedom of expression.

There are flowers with perfume,
There are people with the Grace of Man,
For such flowers and men
That need development, this work is written.

To possess the Grace of Man we must have

Grace of Soul,
Grace of Mind,
Grace of Body.

The Grace of Man is the harmony of
the Grace of Soul, Mind and Body.
To be graceful of only one or two of these
elements, does not make a harmony of the whole.

To dispute the existence of any one of the Trinity of the Soul, Mind and Body, one must deny the existence of the other two.

TO REACH THE SOUL

The Body must be perfected. The mind expresses its intelligence through the motions of the body, until this intelligence calls for a deeper feeling from the heart, which opens the channel, through which flows our Personal Magnetism and

THE SOUL RESPONDS

With its flow of Personal Magnetism or Soul Magnetism; and it is then registered at the brain, the seat of intelligence. Constant cultivation of this nature will develop an intelligence from the Soul which is an inspiration. The mind or mental force, through the action of the body, gives out this soul feeling until you (the student) can feel this soul presence, and it will touch your auditor and he will feel it, and then we are completing the Harmony of the Soul, Mind and Body, which is

THE GRACE OF MAN.

DUALITY.

DUALITY.

1. God and Man.
2. Man and Woman.
3. Day and Night.
4. Light and Shade.
5. Love and Hate.
6. Joy and Sorrow.
7. Spiritual and Physical.

In this study—Duality is in the opposites, and one is not complete without the other.

The Harmony is in the blending of these two by a third element.

1. God and Man—by the Soul.
2. Man and Woman—by Marriage.
3. Day and Night—by Time.
4. Light and Shade—by Emotion.
5. Love and Hate—by Will Power.
6. Joy and Sorrow—by Love.
7. Spiritual and Physical—by Life Force.

NOTE—Hate is essential to Love, on lines of purity. Love is purity. To Love we must Hate impurities.

In this short study the student if he has carefully gone over this work, should be able to understand a volume, as this subject has been covered by most eminent writers as Emerson and others. The author wishes to state that in showing the Duality, the additional philosophies are brought out, viz: That one is essential to the other and why; by showing the third element or link that constitutes a complete harmony and makes a harmony of the whole.

STUDIES IN GRACE.

STUDIES IN GRACE.

The studies given here are exercises, most of which are from the different parts of the study of Curved Lines, blended together making a complete story. They are placed in rotation and are the result or fruit of the work and may be executed in groups, or by an entire class, or as a duet, trio or solo; used as an artistic performance or number on a program. Slow gavotte music is suitable; the one given on the last page applies more to the "Visions or play of the Emotions," although it is suitable for either. The first study, "The Blending Pictures," should be taken up at the seventh lesson; the second study "The Visions," at the twenty-first lesson; the third study "A Solo," at the thirty-fifth lesson, requiring fourteen lessons with the regular study to complete each number, making a total of fifty lessons. Although they may all be mastered in twenty-one lessons.

THE BLENDING PICTURES.

The Blending Pictures consist of seven sets of poses and are representations from works of art, sculpture, paintings, creations of fiction and subjects from daily life. The perfection of their execution is the perfect blending of one pose into that of the other; no break occuring from the opening to the finale. They are executed in the following order per the illustrations and explanations on page 170. (1.) The Gladiator, Niobe, Attraction and Surprise. (2.) The flight of the birds, showing ten Attitudes. (3.) The three statues, the Water Carrier, Grace (the Arch Pose) and Justice. (4.) Sorrow, Divine Supplication, Prayer, Resignation, the Diver, Defiance, the Sword, and the Cause Forever. (5.) The Side View Attitudes, the Pirouette and Coquette (courtsey). (6.) Anticipation, Horror, the Fall, the Awakening. (7.) The Arabesques and finale.

THE VISIONS, OR THE PLAY OF THE EMOTIONS.

This is a recital of seven visions, the subjects seen and described by the interpreter and may be executed by a class standing in couples. When used as a solo, at Visions Nos. 5 and 6, the movements of one person are first executed and in the repetition of the Vision the second persons are impersonated. Each Vision is a story in itself. These Visions are supposed to have appeared to the reciter and are vividly retold as per illustrations and explanations on page 174. Their perfect execution depends greatly on the continuous blending of one pose into that of the other. A short description of each Vision is appended.

THE VISION OF ONE ON HIGH.—Listen, while I relate of a wonderful vision of a celestial being and I an earthly worm, (from page 90, No. 6.)

THE VISION OF THE GATEWAY OF HEAVEN—DEATH.—Emerging through the Veil of Isis came Death; overcome with horror, I faint; awakening, fascination draws me closer, then with exhilarated joy, I recognize Death as but the Gates of Heaven, (from page 80, No. 7.)

THE VISION OF ANGELS AT PLAY.—A blending of Harmony and Grace inspired by a divine choir, with reverence for a sublime power always uppermost; then the blendings end in a picture.

THE VISION OF "IN A GARDEN FAIR."—I am transported to a veritable Garden of Eden. An Apollo gliding toward a balcony whereon stood his goddess; sending forth his whole soul, he attracts her attention. In the exhilaration of bliss he moves on air; a kiss, a heart of joy, ecstasy, seems to spring from his soul, (from page 80, No. 6.)

THE VISION OF THE ACCUSED.—Now Stygian darkness in which stands one accused, beside her the accuser, stern and relentless; she beseeches mercy, then for a moment sympathy vibrates in both hearts; the next attitude is the one of prayer, and that of the other, humility.

THE VISION OF THE MINUET.—The cloud of age is now unfolded, soft music of a bygone era floats through the air. In powdered wig and buckled shoes, with lady fair, they pose the Minuet; the stately turn, they pose in arms, a blending and the salutations—'tis seen no more.

THE VISION OF THE GRACES.—Floating, blending goddesses with garlands round entwined, in procession now appeared. The panel figures of an old mosaic hall were living, breathing beings now once more. Fainter grew the mist—the procession passed from view.

This description is usually printed on the program and sometimes recited before the opening.

A SOLO.

Taken from Exercises in this Book.

In this study the selection has been made of exercises relating to the soul and higher emotions. Each study being complete within itself. There are perfect blending movements at the finish of each, connecting them with the commencement of the one following; this will also be found in the "Blending Pictures" and "The Visions." They are executed in the following order, per the illustrations and explanations on page 178. (1.) "The Call to God." (2.) "The Repentant." (3.) "Take my Soul in Thy Keeping." (4.) "A Pastoral." (5.) The "Blendings." (6.) "The Emotion of Love," and (7.) "The Finale." This entire study relates to the Soul and the Spiral Movements. A description for their proper expression and intent are herewith given.

THE CALL TO GOD.—To Him, Almighty Power, I consecrate my soul, my all to thee. Take me, sinner that I am, I fear my soul; I am unheard, but no, it is the Merciful Shepherd, the Almighty God of all, (from page 86, No. 2.)

THE REPENTANT.—Shadows enshroud me, misery of an unknown fear possesses me. A ray of light, one hope, with outstretched arms I try to grasp, but bitter disappointment. Another chance to save this soul of mine, but blank despair confronts me, a reason lights mine eyes, at last to Thee, O God of all I trust, my prayer e'en now gives me content, I am resigned, (from page 72, No. 6.)

TAKE MY SOUL IN THY KEEPING.—A dream of golden sunshine bursts softly over all, it is the Soul's awakening. The palpitation of my bliss cannot be held. What joy supreme, my soul, my life! O spare this weak and earthly casement. Behold the outer shrine of my Creator, Thy presence doth enwrap me, I am Thine, (from page 88, No. 3.)

A PASTORAL.—In nature's woodland garden, a poet roamed. The air seemed filled with music from a babbling brook and all the forest life. Enwrapped in nature's folds, his soul responded to her calls. From mild interest, all-absorbing the story grew, until the inspiring greatness that surrounded him, inspired him, and carried his ardor to the floating realms above, (from page 78, No. 4.)

THE BENDINGS.—With grace in every blend and turn, the figures move. The attitudes show suppleness and are very graceful, (from Study of Attitudes, page 110.)

THE EMOTION OF LOVE.
"As the stars above could never be more true,
There is my heart, my dearest one for you,
I'd love to tell the story o'er and o'er,
And in thy presence be for ever more."—*Song.*
(from page 78, No. 5.)

THE FINALE.—With ease and grace personified, their bodies poised in balancing attitudes, as if suspended in mid air. The god Mercury now appears; a graceful curve or turn, and then these graceful gods have disappeared.

The Blending Pictures.

No. 1. Poses. Gladiator, Niobe, Attraction, Surprise.

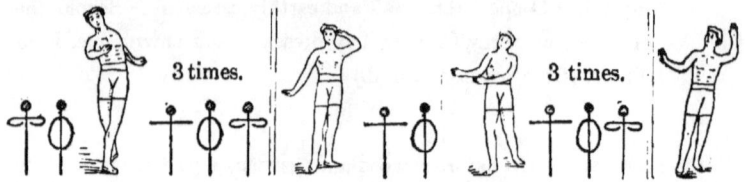

No. 2. The Flight of the Birds.

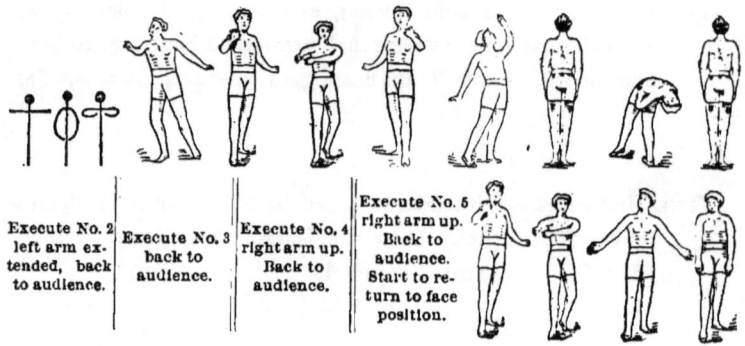

No. 3. The Statues and Water Carrier.

Arch Posing. Justice.

EXPLANATION.

Executed by a group—stationary; standing in separated positions. Use (soft,) slow gavotte music with expression; to assume attitudes count 4, and hold attitude, 4 counts.

No. 1. GLADIATOR, (3 times) AND NIOBE. ATTRACTION, (3 times) AND SURPRISE.

Start by passing arms through positions and assume pose of Gladiator per cut 1; this is a Straight (forward knee bend) Attitude, with left knee bent, right arm in front and left arm in counter pose. Dissolve attitude to natural position and repeat Gladiator (reverse) by sliding left foot forward, etc. Dissolve attitude to natural position and assume Gladiator as at first, moving forward at each pose. The fourth time (pass arms through arm positions) assume pose of Niobe, per cut 2. Body natural (pass arms through position) and assume pose of Attraction, per cut 3; dissolve attitude to natural position and repeat Attraction to left, then again to right; the fourth time (arms through positions) assume pose of Surprise, per cut 4.

No. 2. THE FLIGHT OF THE BIRDS.

Indicate the start of the flight by pointing, per cut 1. The flight is traced to front, per cut 2. Cuts Nos. 3 and 4 show where the trace is taken up by the left hand. Cut No. 5 illustrating their flight, going around and overhead, the head and hand following, until the body is turned about, per cut 6, the back bend now occurs, per cut 7. Cut No. 8 is recovering to an upright position; now the same movements, per cuts Nos. 2, 3, 4, 5, are repeated with back to audience; this last is the body turning (face) front. Cut 9 places arms same as second pose; cut 10 is arms preparatory to assume first position. Cut 11 is a slow fluttering movement, denoting the birds' settling where they started from. Cut 12 showing the finish. Every movement is illustrated, and should be executed as one continuous movement; the feet are moved very little; the body being turned by pivoting on the ball of the feet. This number requires 16 bars of music.

No. 3. THE STATUES (3 times) AND WATER CARRIER. THE ARCH POSE (2 times) AND JUSTICE.

Pass arms through positions and assume first statue; this is an Arch (forward right) Attitude, ¾ view front, per cut 1. Dissolve to a natural position, repeat same to forward left and assume second statue, per cut 2. Dissolve attitude and step back with left foot, (pass arms through positions) and assume third attitude; this is a Straight (back knee bend) Attitude. Dissolve to a natural position and assume Pirouette pose, per cut No. 4, turn and assume pose of Water Carrier, per cut 5. From pose of Water Carrier form Arch (right) Attitude, per cut 6. Continue arm movements to cut 7; continue arm movements to cut 8; continue arm movements to cut 9; then continue arm movements to cut 10, making a turn and finish by assuming the pose of Justice, per cut 11.

No. 4. SIDE VIEW ATTITUDES. PIROUETTE AND COQUETTE.

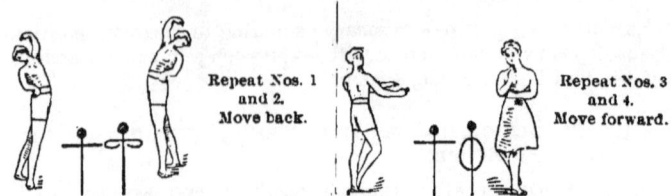

No. 5. POSES. SORROW, DIVINE SUPPLICATION, PRAYER, RESIGNATION, DIVER, DEFIANCE, THE SWORD AND THE CAUSE FOREVER.

No. 6. POSES. ANTICIPATION, HORROR, FALL, AWAKENING.

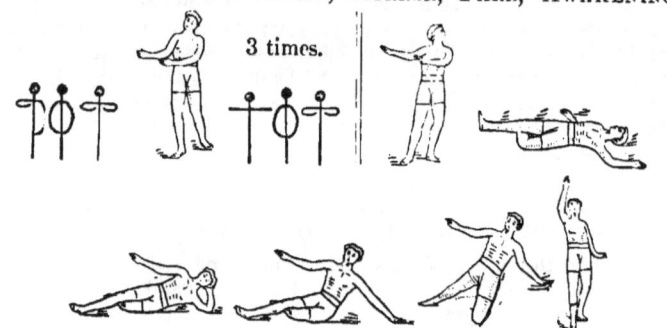

No. 7. THE ARABESQUES AND SALUTES.

No. 4. SIDE VIEW ATTITUDES (4 times.) PIROUETTE AND COQUETTE (4 times.)

From the pose of Justice to the Side View Attitudes, it is only necessary to step back with left foot, presenting the right side view of the body; the arms are already in position, per cut 1. Follow arm positions and place right foot behind and present a left side view of body, per cut 2. Repeat right side, as per cut 1; repeat left side, per cut 2; move back at each attitude. Now pass arms through positions, step forward and assume pose of Pirouette and turn, per cut 3; continue arm positions and do pose Coquette, per cut 4. Execute this Pirouette, turn and Coquette (moving forward) 4 times.

No. 5. POSES. SORROW, DIVINE SUPPLICATION, PRAYER, RESIGNATION, DIVER, DEFIANCE, THE SWORD, THE CAUSE FOREVER.

Start by assuming natural attitude; then blend to pose of Sorrow, per cut 1. Dissolve to natural position, follow arm positions and assume pose of Divine Supplication, per cut 2; this is an Arch (outward) Attitude. Next—this same attitude, place arms across breast, per cut 3. Slowly kneel, per cut 4 and bend back. Slowly bend forward and relax, per cut 5. Rise, pass arms through positions and assume pose of the Diver, per cut 6; this is an Arch (inward) Attitude. Dissolve to natural position, follow arm positions and assume Arch (outward) Attitude, per cut 7; leave arms down to crossed on breast and assume a Straight (back) Attitude, side view, which is the pose of Defiance, per cut 8. Follow arm positions and assume pose The Sword; this is a Straight (forward knee bend) Attitude, per cut 9. Dissolve arms (retain same attitude,) per cut 10; at last pass arms through positions and spiral the body and arms, per pose, The Cause Forever, cut 11.

No. 6. POSES. ANTICIPATION (3 times) HORROR. THE FALL AND THE AWAKENING.

Start to assume natural position from last attitude, then assume pose of Anticipation to right, per cut 1; follow arm movements and repeat same to left; repeat to right. Dissolve to natural position, follow arm movements and assume pose of Horror; this is a Straight (left knee bend) Attitude, per cut 2. At count of four of that bar of music *fall*, per cut 3; remain thus to end of strain of music. Now slowly rise $\frac{1}{4}$, per cut 4; the hand traversing face as if Awakening; now rise $\frac{1}{2}$, per cut 5, and repeat hand movement. Next kneel, per cut 6, and start the spiral movement of body and arms—rise and finish, per cut 7.

No. 7. THE ARABESQUES AND SALUTES!

Pass arms through positions and assume Pirouette position and turn, per cut 1; when turning pass arms through position and at face front assume an Arabesque attitude, per cut 2. Repeat the Pirouette (turn), per cut 3, and when face front Salute, per cut 4. Repeat this number until exit or to the finish of the strain of music.

NOTE—Move backward on each pirouette, count 4 for a movement and 4 for an attitude.

The Visions or Play of the Emotions.

No. 1. The Vision of One on High.

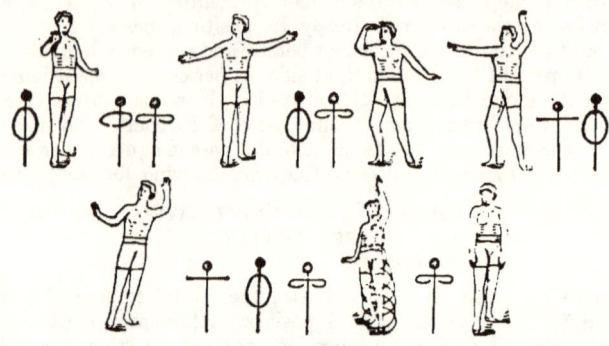

No. 2. The Vision of the Gateway of Heaven—Death.

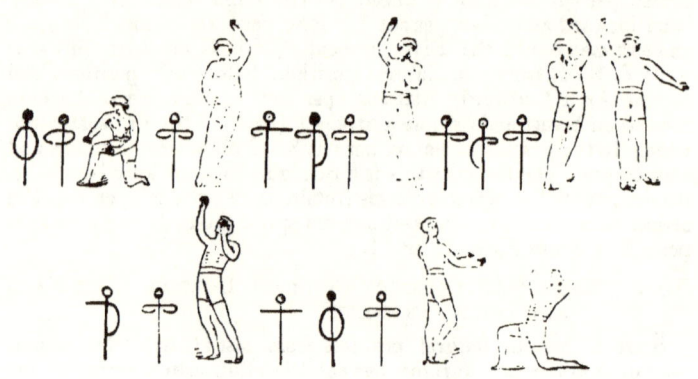

No. 3. The Vision of the Angels at Play.

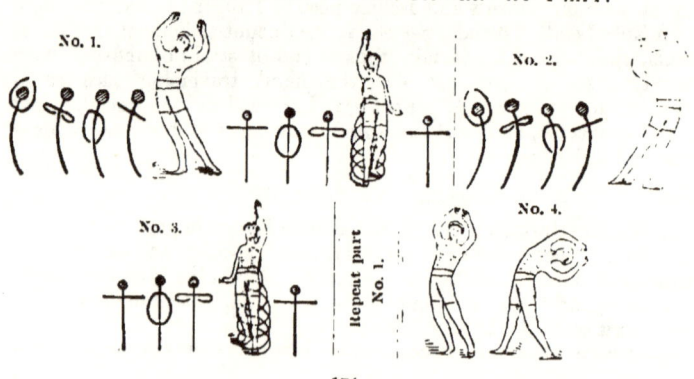

EXPLANATION.

Executed by a group—stationary; standing in couples. Use Viola Gavotte, page 182. Count 4 each to assume attitudes, and to hold attitudes.

No. 1. THE VISION OF ONE ON HIGH.

Pass arms through positions and assume first pose, "Attention," per cut 1; this is a Straight (forward) Attitude. Follow arm movements and assume pose "Telling a Multitude," per cut 2; this is a Straight (back) Attitude. Follow arm movements and assume pose "There Appeared," per cut 3; this is a Straight (right knee bend) Attitude, and point right hand, left going to 1st upper, assume pose "Was Seen," per cut 4. Follow arm movements and assume pose, "Surprise," per cut 5; this is a Straight (back knee bend) Attitude. Recover to natural position and execute a Spiral Motion of the body (left hand leading) and assume pose "Of One on High," per cut 6. The last is a collapse or relaxation, per cut 7. Repeat entire (reverse) start with left arm, etc.

No. 2. THE VISION OF THE GATEWAY OF HEAVEN.—DEATH.

From standing position move arms and slowly kneel (right foot in front) and assume pose, "Discovery," per cut 1. Rise slowly, move arms and assume pose, "Unable to Fathom," per cut 2; this is a Straight (back knee bend) Attitude. Follow (body natural) arm positions, blend body and assume pose, per cut 3; this is an Arch (left knee bend) Attitude. Now (body natural) move arms and assume pose, per cut 4; this is an Arch (right knee bend) Attitude; the expression of the last two attitudes means "Curiosity and Dawning," now step on right foot and assume pose, per cut 5; this is a Straight (forward knee bend) ¾ front; clinch hands; the expression of head and shoulders is "Horror." Remain with weight on right foot, assume arm position and assume pose, "Realization," per cut 6; this is an Arch (outward) Attitude, ¾ front view. The last is Pirouette, turning as if fainting, per cut 7; now kneel, per cut 8; meaning "The Truth—Heaven." Rise and repeat entire, reversing all movements. The mental is now purely spiritual and the face should express Seeing the Gates of Heaven.

No. 3. THE VISION OF THE ANGELS AT PLAY.

From last attitude simply rise and assume pose, per cut 1; this is an Arch (right) Attitude; dissolve to natural position, follow arm movements and execute Spiral Motion, per cut 2. Now assume 3d position of arms and blend to an Arch (left) Attitude and repeat same on left side, per cuts 3 and 4. Now assume 3d position of arms and repeat same to right side, per cuts Nos. 1 and 2. Finish by placing both hands over head, pivot or turn around, per cut 5; at face front bend, per cut 6.

No. 4. THE VISION OF IN A GARDEN FAIR.

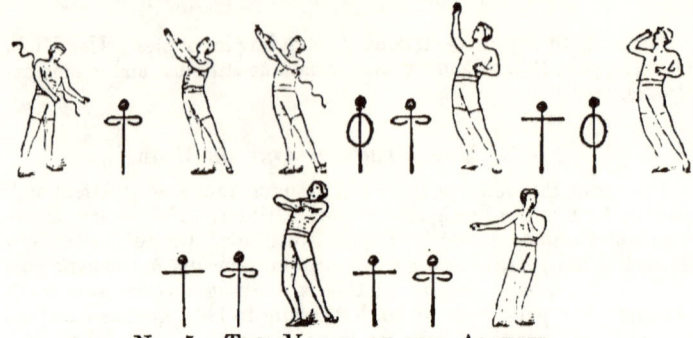

No. 5. THE VISION OF THE ACCUSED.

No. 6. THE VISION OF THE MINUET.

No. 7. THE VISIONS OF THE GRACES. THE ARABESQUES.

FINALE.

No. 4. The Vision of In a Garden Fair.

From last attitude move body gracefully forward and assume pose, per cut 1; now move both arms from left side diagonally up to right side; assume pose, "He Calls to Her," per cut 2. Return arms to left side look up, and assume pose, "His Presence There," per cut 3. Then (body natural) follow arm movements and assume pose, "His Heart," per cut 4; this is an Arch (right) Attitude, ¾ front view, with left hand on heart. Follow arm movements (body natural) and assume pose, "His Lips," per cut 5; the fingers traverse lips. Follow (body natural) arm movements and assume pose, "His Worship," per cut 6; this is an Arch (right) Attitude and hands are clasped. Now move arms and assume pose, "His Perplexity," per cut 7; this is a Straight (left knee bend) Attitude. Repeat entire from right side, use reverse positions. Use Trio of Gavotte.

No. 5. The Vision of the Accused.

Both, pass arms through position and assume attitude, per cut 1; the first is "Weeping," which is an Arch (inward) Attitude, ¾ view front; the second is the "Accuser." Dissolve to body natural and assume attitudes, per cut 2; the first is an Arch (outward) Attitude; the second is a Straight (back) Attitude. Dissolve attitudes assume next picture, per cut 3; these are both Straight (knee bend) Attitudes. Dissolve attitudes and assume last picture, per cut 4; the first is a Kneeling Attitude; the second, the arm movement of left arm traverses the brow before assuming the attitude. Repeat entire.

No. 6. The Vision of the Minuet.

Dissolve to a natural position, then both give right hand to partner, raise it and extend right foot, per cut 1. Now lady turns backward, per cut 2, and moves backward into partner's arms, per cut 3. Next dissolve attitudes, which places them in their partner's place; turn slowly around, per cut 4; when face front, salute, per cut 5. Repeat entire in reverse by giving left hand and pointing left foot, etc.

No. 7. Vision of the Graces. The Arabesques.

Assume arm positions, with an Arch (left side) Attitude, follow arm movements per cuts; at 3d position body blending through natural body pose to an Arch (right side) Attitude, and assume Mercury pose, per cut 1. Continue with body in this Arch Attitude (the face to front) and pass arms through positions per cuts until 3d position, here blend body through natural pose to an Arch (left side) Attitude, and assume Mercury Pose, per cut 2. To continue from last Attitude the lower hand is passed to 1st upper as per cut; from this position return to beginning, repeat entire, making one continuous blending.

Finale.

Assume Pirouette and turn, same time moving backward, per cut 1, at face front salute per cut 2. Repeat Pirouette (reverse) turn, going backward, per cut 3; at face front salute, per cut 4. Repeat until off stage or to the end of the strain of music.

A Solo, from this Study.

No. 1. The Call to God.

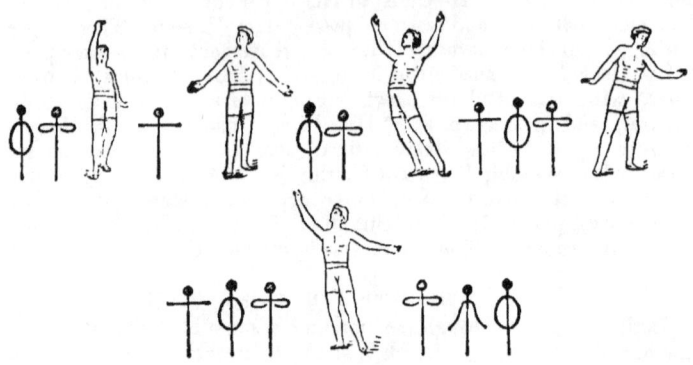

No. 2. The Repentant.

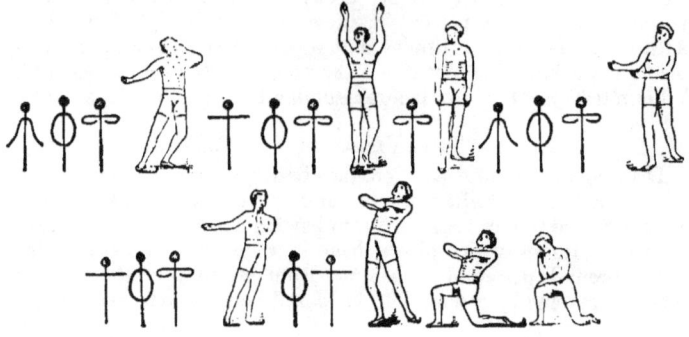

No. 3. Take my Soul in Thy Keeping.

EXPLANATION.

Executed as a Solo, but may be performed by any number. Stationary positions, well separated. Use slow Gavotte music or an appropriate Hymn, and sway to its rhythm.

No. 1. The Call to God.

Execute a spiral movement of the body and assume pose "To Him," per cut 1. Move arms through next positions, and assume a Straight (back knee bend) Attitude; look up as if "Offering yourself to God," per cut 2. Follow arm movements, (body natural,) and by a spiral body motion assume pose "The Call to God," per cut 3. This is an Arch (outward) Attitude, ¾ view front, and is easily assumed by placing weight on front foot. Move arms through positions (body natural) and assume pose "The Doubt," per cut 4. This is a Straight (left side knee bend) Attitude, the left hand traverses face from second position. Follow arm movements, and assume pose "Answered," per cut 5. Recover a natural pose by following the last positions. Repeat entire (reverse) left hand and actions to left, etc.

No. 2. The Repentant.

Move arms to first impersonation "The Penitent," per cut 1. This is an Arch (inward knee bend) Attitude; the arms go to head and 3d lower position, with the entire body and mentality expressing the sense of the subject. Next, (body natural,) pass the arms through positions and assume pose "The Cry of Anguish," per cut 2; the physical and mental expressing the sense, by starting from a slight sinking movement, and thus emphasizing a stronger effect. Follow arm positions, and assume pose "Hope Gone," per cut 3. The drooping of the head and an inward bend well express it, if the mind has been so imbued. Follow arm movement and assume pose "A Faint Hope," per cut 4. This is a Straight (right knee bend) Attitude, the inclination of head denoting the intent. Follow arm movement and assume pose "Disappointment," per cut 5. This is a Straight (left) Attitude, hand to chin and right extended. Follow arm movements and assume pose "To God I Trust," per cut 6. This is an Arch (outward) Attitude, with hands clasped; now simply kneel, per cut 7, leaning backward. The last is a relaxation study forward, per cut 8. Rise and repeat (reverse), right hand and actions to left, etc.

No. 3. Take My Soul in Thy Keeping.

From kneeling pose, rise and execute a spiral motion of the body with arm movements, and assume pose "The Soul's Awakening," per cut 1. This is an Arch (outward) Attitude, ¾ front view, facing right. Recover, (body natural), and at same time about face (to left); follow arm movements and assume pose "Suppressed Suspense," per cut 2; covering the face, swaying the body back and forth. Recover, (body natural), execute a spiral motion, and assume pose "My Soul," per cut 3. This is an Arch (outward) Attitude, ¾ front view, facing left. Now slowly slide right foot backward and kneel, (eyes looking up), "In Thy Keeping," per cut 4. Repeat, by rising and blending in the first Attitude, face left, etc.

No. 4. A Pastoral.

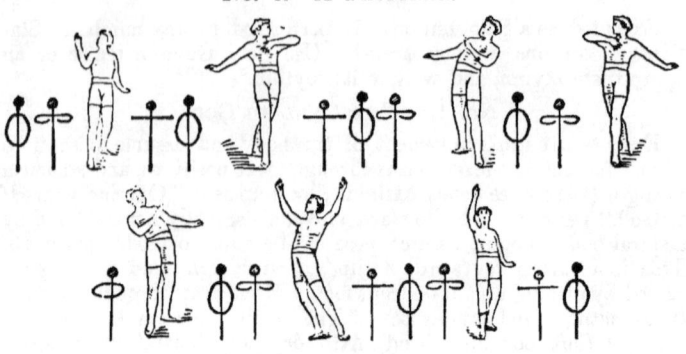

No. 5. The Bendings.

No. 6. The Emotion of Love.

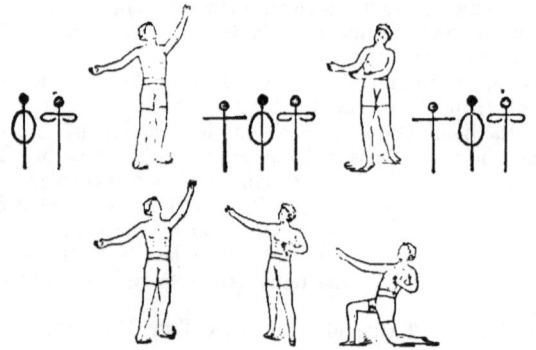

No. 7. The Finale.

No. 4. A Pastoral.

Follow arm positions and assume pose per cut 1; this is a Straight (forward knee bend) Attitude; the Wave Motion is diagonally upward in front; start from a sinking movement of the body and go upward, diagonally forward. Follow arm motions and assume pose per cut 2; this is a Straight (back) pose. Follow arm positions, and assume pose per cut 3; this is a Straight (forward knee bend) Attitude; sway the body forward and backward. Follow arm positions, and assume pose per cut 4; this is a straight (back) pose. Follow arm positions, and assume pose per cut 5; this is a Straight (forward right knee bend) Attitude; sway the body forward and backward. These movements mean "A Poet Wandering," and "Gathering Inspiration." Follow arm positions, and assume pose "Receiving Inspiration," per cut 6; this is an Arch (outward) Attitude. Follow arm positions, and assume pose "I am Inspired," per cut 7; this is a spiral motion of the body. Repeat entire (reverse) left arm and action to left.

No. 5. The Bendings.

Assume Pirouette pose and turn, per cut 1; at face front bend to right, per cut 2. Recover and assume Pirouette pose; turn, per cut 3; at face front bend to left, per cut 4. Recover, assume Pirouette pose, turn ½, per cut 5; when back is to audience, bend backward, per cut 6. Recover and pose side view, per cut 7. Recover and face front and salute, per cut 8. (Not repeated.)

No. 6. The Emotion of Love.

Follow arm positions and assume pose "The Stars Above," per cut 1. Follow arm positions (face right) and assume pose "To Thee, my Heart," per cut 2. Follow arm positions and assume pose per cut 3. Follow arm positions (face right) and assume pose per cut 4. Now kneel, place hand at heart and right extended, as in the pose "At Thy Feet," per cut 5. (Not repeated.)

No. 7. Finale.

Follow arm movements and assume pose "Mercury Balance," to the right, per cut 1. Follow arm movements and assume pose "Mercury Balance," to left, per cut 2. Follow arm movements and assume "Pirouette," and turn, per cut 3; at face front salute, per cut 4. Repeat entire until end of strain, or until off. Note.—At each turn move to right, and when near edge of stage, disappear.

VIOLA
Gavotte.

MAGGIE GOLDEN

Complete copy of this beautiful Gavotte may be had of the publishers,
THE GEO. B. JENNINGS Co., Cincinnati, Ohio.

APPENDIX.

For the Teacher, (or Self-Improvement.)

Each study is complete under separate headings, starting with easy construction for No. 1, and increasing to construction No. 7. The student or teacher should read the book first, and then master each part complete.

To Impart the Study.

The lesson hour should be divided into periods, per the following classification. The number of lessons necessary should be left to the teacher. Herewith are given the periods of advancement to impart and master the work; these periods simply show how the work is mastered. At the seventh class lesson, the first Study in Grace should be started. Fifty lessons of one hour each, complete the work.

1st Period.
Physical Corrections, ¾ time.
Grace Lines .... ¼ "

2d Period.
Physical Corrections, ½ time.
Grace Lines .... ¼ "
Construction .... ¼ "

3d Period.
Physical Corrections, ½ time.
Grace Lines and Construction .... ⅜ "
Philosophy .... ⅛ "

4th Period.
Physical Corrections .... ½ time.
Grace Lines and Construction ...... ¼ "
Philosophy and Force .... ¼ "

5th Period.
Physical Corrections .... ⅜ time.
Grace Lines and Philosophy ...... ¼ "
Force and Harmony .... ¼ "

6th Period.
Physical Corrections .... ½ time.
Grace Lines and Force .... ¼ "
Harmony .... ¼ "

7th Period.
Physical Corrections .... ½ time.
Force and Harmony .... ⅛ "
Studies in Grace .... ⅜ "

Classifications.

1. **Physical Corrections.**—Include the entire study of Gravity, pages 7-16; Physical Corrections, Series I, II, III, pages 17-51; and Flexibility, pages 95-102.
2. **Grace Lines.**—Includes the entire study of Curved Lines, 63-93.
3. **Construction.**—Includes the entire study of Attitudes, 105-120.
4. **Philosophy.**—Includes the entire study of the Anatomy of the Elements of Man, pages 53-61; the study of Mind, Inspiration, Ideality, Duality, pages 151-164.
5. **Force.**—Includes the entire Study of Force, (including Walking,) pages 121-134.
6. **Harmony.**—Includes the entire study of Poise, Harmony and Opposition, pages 135-149.
7. **Grace.**—Includes the three studies in Grace, pages 165-183.

Sub-Classification.

No. 1. Physical Corrections:
1. Gravity .... page 7.
2. Series I ..... " 17.
3. Series II ..... " 29.
4. Series III ..... " 41.
5. Flexing .... " 95.
6. Relaxation .... " 100.
7. Falling .... " 102.

www.ingramcontent.com/pod-product-compliance
Lightning Source LLC
Chambersburg PA
CBHW020839160426
43192CB00007B/717